Horse Breeding

is to be returned on or before
last date stamped below.

3

3

HORSE BREEDING
A PRACTICAL GUIDE FOR OWNERS

— Tony Pavord and Marcy Drummond —

The Crowood Press

First published in 1990 by
The Crowood Press
Ramsbury, Marlborough
Wiltshire SN8 2HE

British Library Cataloguing in Publication Data
Pavord, Tony
  Horse breeding.
  1. Livestock: Horses. Breeding. Stud farms
  I. Title II. Drummond, Marcy
  636.1'082

ISBN 1 85223 170 X

Photographs by Tony Pavord and Marcy Drummond
Line-drawings by Elaine How

Typeset by Alacrity Phototypesetters, Banwell Castle,
Weston-super-Mare
Printed in Great Britain at The Bath Press

Contents

Acknowledgements

The authors acknowledge, with thanks, the assistance of many horse owners, studs and breed societies who provided information and subjects for photographs, in particular Hilary Morgan, Sue Richards, the Pembridge Stud, Bru Marba Stud, Liz McCurley, Bridget Barnes, Val Long, Becky Beaumont and Jane Evans.

1 The Decision to Breed a Horse

Breeding and training your own horse can be the most rewarding of equestrian experiences, provided you know, before you start, that any horse you breed will be happy and well cared for in his future life. If you are sure that, with the right stallion, your mare will produce an offspring that you will rear, keep and use yourself, then go ahead. Otherwise, think long and hard before you breed a foal.

The world is full of unwanted horses and ponies which, due to random breeding and bad management, are unsuitable for any useful purpose. A few lucky ones find homes in welfare sanctuaries, dedicated to their care. The rest – the vast majority – are either neglected or sold on, time and again, until, at last, broken down and unloved, they are herded from the sale ring into the meat man's lorry and thence to the abbatoir. In this supposedly enlightened age, no equine should be born to such a fate. If you breed a horse or pony, he will need care and attention – not just for the first few months, when he is a fluffy and appealing foal, but while he is growing up and for the rest of his life. A boisterous, growing youngster needs correct handling and knowledgeable management if he is to realise his full potential. As he matures, he will be trained to carry a rider, or 'broken in' and will be expected to work for a living. A horse bred without athletic potential, with poor conformation, for example, or one who is neglected during the early years, so that his development is curtailed, will be at a disadvantage when asked to work hard later. These are the horses who join the sad rows of animals waiting to be sold at markets.

Although many markets today are adequately run and warranties are often given as to the animals' soundness, they are still often the last refuge of the seller who has been unable to find a private buyer. A horse may be being sold because he frequently goes lame for no apparently serious reason; perhaps he has a dust allergy or is a 'roarer'; perhaps there is a temperament problem.

Buyers at markets are often on the look-out for a cheap bargain. They have little to lose and when a horse proves unsuitable, back he goes again to the sale ring. These might seem depressing thoughts at the beginning of a book about horse breeding, but they emphasise the need to make an informed decision. An inexperienced owner who wants to breed a foal should seek expert help, as regards to choosing a stallion and to what is entailed in caring for the pregnant mare, the new foal and the developing youngster. 'Look before you leap' is as good a maxim in breeding as in any other sphere of equestrianism!

The usual reason for the private horse owner deciding to breed a foal is the desire to own a horse from birth and to build up a close bond of trust and com-

Fig 1   Breeding your own foal – the most rewarding of experiences.

panionship with him. This is an attractive prospect, but it is essential to be sure that your mare will breed the type of horse that will suit your needs. After all, the mare represents 50 per cent of the equation that produces the foal. The classic example of the wrong attitude towards breeding is the decision to breed from a mare who has been retired early from normal work due to some disability. The owner sees her standing idle in the field and thinks he might as well breed a foal, on the grounds that it will not cost much and at least he can sell it to pay for the mare's keep.

A mare who has been retired due to an injury that was no fault of her own may well breed a useful foal, but if, as is quite likely, retirement was enforced due to some inherent defect or weakness, this could well be passed on to any offspring. Such a mare is not a good breeding prospect. Another frequently made mistake is the mare being sent to a stallion who happens to be standing just down the road because it is convenient and cheaper than sending her away to stud. A stallion should be chosen on his merits as a sire in conjunction with the mare concerned and not just because he is close at hand. A foal born with the worst features of two mismatched parents, plus some inherited defect, is of little use to anyone.

Many owners would like to breed

Fig 2   *This Thoroughbred/Irish Draught mare, Zeta, would be an ideal breeding prospect, crossed back to a Thoroughbred stallion, or to an Arab, Anglo-Arab, or Warmblood, to produce a quality competition or riding horse. Her conformation is good and she has proved her stamina and soundness in long-distance competitions.*

from their own good mares, but are reluctant to take a performance horse out of work for a year or more to produce a foal. However, modern technology has taken care of that, as it is now possible to produce foals by embryo transplant. The genetic dam continues in work, while a surrogate mother carries and rears the foal (*see* Chapter 14).

Once you have decided that there is no physical reason why you should not breed from your mare, take a critical look at her and decide what kind of foal she would be likely to breed. Take into account what you want to do with the eventual offspring. Is he to be a show jumper, or a dressage horse; a show hunter or a driving horse? Even if you just want a general-purpose 'fun' horse, your choice of sire will have a considerable effect on your youngster's type and temperament.

Most riding horses are not pure-breds. In Britain they come mainly from that extensive and valuable pool of equine stock loosely termed 'hunters'. The hunter is not, of course, a breed in its own right, but a type of horse having attri-

9

butes which make it suitable for general-purpose riding, and hunting in particular, and which is drawn from many breeds.

The hunter's original purpose, which might seem to state the obvious, was to carry a rider in the hunting field. For this he needed both strength and stamina and, in the more renowned jumping countries, a considerable amount of courage and athletic ability. At the more elegant end of the scale is the lightweight 'ladies' hunter, whilst in the heavyweight division a horse must be up to carrying a rider of 14 stone or more.

Showing classes for hunters stemmed from the hunting field, as did the sports of point-to-pointing and, more recently, team chasing, which are both immensely popular today. Although top show horses and Thoroughbred point-to-pointers may put in very limited appearances in the hunting field nowadays, it is to those sound, strong, genuine horses who could 'go all day' that they owe their origin.

Similarly, the vast majority of British riding horses and hacks are the result of hunter breeding policies, many being at least part Thoroughbred, with a huge variety of other breeds being added to the

Fig 3   Hunters need to be sound, forward going horses, with good bone and stamina. An equable and obedient temperament is necessary to cope with crowded meets and the excitement of hunting. The hunting tradition formed the basis of Britain's pool of breeding stock for both show and competition horses.

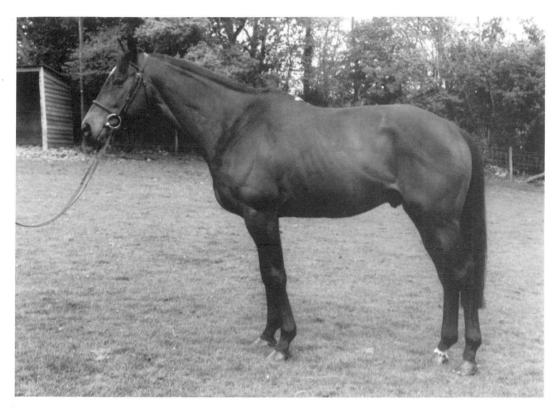

Fig 4   The 'straight' Thoroughbred has become increasingly used as a competition horse. This seven-year-old, by the stallion, Free State, is successfully competing in horse trials. His conformation is short coupled and compact, but athletic.

melting pot. The influence of the Thoroughbred in improving the quality of riding horses cannot be denied, although it has also caused problems when Thoroughbred sires have been used non-selectively (*see* Chapter 2).

The Thoroughbred is known as a 'hot blooded' breed, bestowing qualities of spirit, speed and stamina on other breeds when used as a cross. Another hot-blooded breed, the Arabian, is also used to produce quality horses, usually of a smaller, finer type than the Thoroughbred cross. Many of these become show

hacks or small show hunters, but they can also make superb competition horses within the limits of their size.

For many people, the top show hunter is the nearest possible example of equine perfection. Top show horse producers know only too well that such horses are extremely difficult to find. A true champion will stand out above his fellows and the show ring itself is a limiting factor – there are far more people anxious to win prizes than there are prizes to be won. A horse with exceptional show potential will always realise a high price, if the

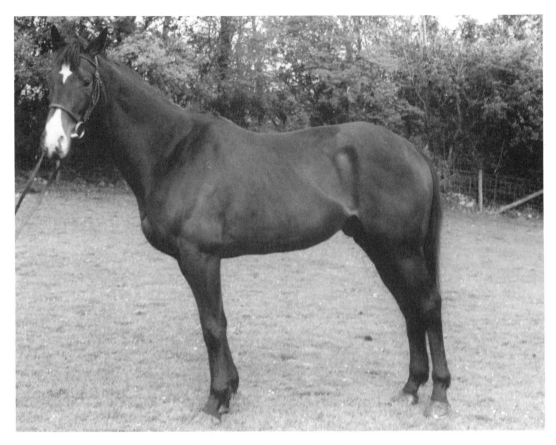

Fig 5 *His five-year-old half-brother, also by Free State, has the rangier conformation more usual in the breed. He is also destined for a career in eventing.*

seller is aware of its worth, but to breed such a sought after animal needs a great deal of luck as well as sound judgement, even for experienced horse breeders. 'Handsome is as handsome does' is another saying often heard around the show ring, implying that show horses might look pretty but are not capable of doing much else. This is often unjustified, as many good show horses go on to do well in other spheres – eventing, show jumping and hunting.

Over the past twenty years, however, showing has developed an unfortunate fashion for horses to be shown rather too

well covered with flesh. This may be termed 'show condition' or 'fat', depending upon your point of view, but an excess of it can result in later joint problems for the horses, especially where growing youngstock is concerned. It is important to understand that beneath the layers of flesh, the features which make a show horse attractive are the same features which provide athletic ability and scope for performance in a different role. Even in the show horse, the ability to move well at all paces, including a good gallop, must be demonstrated.

The essence of this is that in the judge's

eye a horse's beauty should be directly related to his ability to perform. To put it another way, anything which is well designed is both attractive to look at and efficient in performance. So how do we analyse that beauty, in terms of athletic potential? There are three components to consider – conformation, paces and temperament. The first two can be linked, as a horse with good conformation will often have good paces. Good conformation can be summed up by the overall image presented by a horse. As one top breeder puts it, 'A horse should be the kind of picture you want to look at over your mantelpiece every day'.

A horse should stand four square, i.e. with 'one leg at each corner', and each part of his body should be in proportion with the rest. The head, containing the dense bones of the skull, is proportionately heavy for its size and should not be too large, since the horse uses his head and neck to help him balance and a head that is too heavy will make this more difficult. However, in some breeds, where small, 'pretty' heads are considered desirable, efforts to produce them have sometimes led to loss of 'bone' and consequently weak limbs.

A fine head, with well-defined bone structure, denotes quality. Look for a neat muzzle with a well-formed mouth (essential for both chewing up food and accepting a bit), and large nostrils capable of dilating to increase the intake of air when working hard. The forehead should be broad and flat (slightly dished in some breeds), with eyes large, bright and set wide apart.

The ears, which signify the horse's mood as well as enabling him to hear all round, should be mobile and alert; their size will vary according to breed and genetic inheritance. Some owners set great store by large ears, whilst some breed standards, conversely, require small ears for their particular breeds. Some horses even have lop ears, which also endear them to their owners and are supposed to imply a placid and amenable disposition.

The jaw should be clean-cut, with room for the width of a fist between the cheek-bones, to allow plenty of space for breathing and to make it easier for the horse to flex. A prominent windpipe and clearly defined jugular groove also contribute to this.

The head should be neatly attached to a proportionately long and graceful neck, neither short and 'cresty', nor an upside-down 'ewe neck', both of which make the horse more difficult to school and affect natural balance. The neck should flow smoothly down to well-defined withers and shoulders and it is worth mentioning here that although an imaginary picture of a superb specimen of horseflesh would show him well rounded, with strongly developed muscles, he needs the basic framework of a strong and roomy skeletal structure to which that musculature can attach. In this respect, all the angles and joints of the horse's anatomy should be distinctly formed.

The shoulder should be long and sloping, ideally at an angle of 45 degrees. The forearm should be comparatively long and muscular, the knee large and flat and the cannon bone short and strong. The hindquarters should be muscular, with a good length from the point of the hip to the buttock and without sloping away too steeply from the croup. There should be plenty of scope for movement at the stifle and a long gaskin (or second thigh), which runs from the point of the stifle to the hock, with room for good muscle development. The rear of the thigh should curve gently down to the point of the hock.

The hock itself is an important joint containing many bones which are subjected to considerable stress in the working horse. It, therefore, needs to be large, open, strong and flat in front with no lumps or bumps that are not part of good skeletal conformation. When seen from the side, an imaginary vertical line from the point of the buttock should touch the point of the hock and run straight down the back of the leg to the ground. Seen from behind the hind legs should be straight with the hocks turning neither inwards (cow hocks), nor outwards, but directly backwards.

A horse's lower limbs should be 'clean', though hard worked horses may have wind-galls and bumps acquired through injury or trauma. In such cases it may be necessary to make a decision whether this is due to overwork, accidental injury, mismanagement, or a hereditary weakness, before using the horse for breeding. In any case, the cannon bones should be short, flat in front and strong. Much is made of the importance of good 'bone', with particular reference to the measurement around the cannon bone just below the knee. In comparative terms, the greater this measurement and the shorter the cannon bone, the greater the weight the horse can carry. However, not all horses are required to carry great weights and of more importance is the shape of the lower leg.

The cannon bone should run straight down from the front of the knee joint. If the knee joint is further forward the horse is said to be over at the knee. Although this is technically a fault, it does not, in fact, constitute a loss of strength and does reduce strain on the tendons. Lower forelimbs which are narrower at the knee than lower down are referred to as being 'tied in at the knee' and this is a serious weakness as it reduces the space available for the attachment of the tendons and ligaments, which are responsible for the movement of the lower joints. There are no muscles in the horse's lower leg to aid movement, so the relatively inelastic tendons and ligaments must be strong enough to take considerable strain. A leg which curves backwards – 'back at the knee' – is also subject to excessive strain on ligaments and joints.

The pasterns are the horse's shock absorbers. The longer the pastern, the greater the absorption, and the more comfortable the ride, but the greater the strain on the tendons of the leg. Short, upright pasterns give a jarring ride and are less good at absorbing concussion. Upright pasterns usually go with an upright shoulder and a consequently shortened stride. Pasterns should, therefore, be neither too long nor too short. They should have the same angle to the ground as the front of the hoof, and the line made by the front of the hoof and the pastern is known as the hoof/pastern axis. In the front feet, this angle will, ideally, be 45 degrees or slightly more and in the hind feet, up to about 60 degrees, these being slightly more upright than the forefeet.

Good feet are essential for a successful working horse, but it is surprising how often this factor is ignored by breeders. It is true that many foot problems stem from inadequate and infrequent trimming and shoeing, but certain types of feet are more disposed to problems than others.

The greatest problem in Britain is the

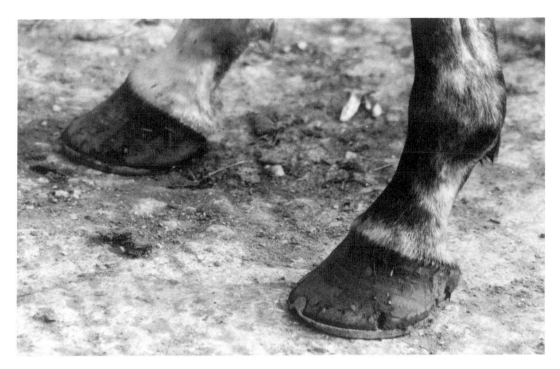

Fig 6   Flat, shallow feet with thin soles and weak hoof walls are a common problem. Correct shoeing can often overcome complications caused by neglect, but breeding from horses with persistently poor feet should be discouraged.

shallow, flat foot nearly always associated with Thoroughbred blood, but the upright, boxy, contracted foot can bring an equal share of trouble. Breeders should look for a compromise – well-shaped, symmetrical feet, with the forefeet slightly rounder than the hind feet. The horn of the hoof wall should be hard, dense and evenly growing, without rings. There should be no cracking nor flaking, nor irregularities of shape. The frog, which helps expand the foot and reduce concussion, should be healthy and well defined. The bars, which help support the heels, should also be clearly delineated, with the heels open for expansion and the sole slightly concave. Poor foot conformation is invariably passed on

to the next generation, although cross-breeding to a breed known to have good feet, for example the Arab, may effect an improvement.

The horse's body is designed to carry the suspended weight of his vital organs, but we insist on his back being adapted to carry the compressed weight of a rider. This requires considerable muscular development, produced by schooling, and, to be successful, basic conformation must lend itself to this purpose. A short, compact back is obviously a strong back, whilst a longer back indicates the potential for speed and provides a smooth ride. A long back is acceptable, particularly in a mare, who may need room to breed foals, provided the length is distributed over the whole back and does not just result in over-long weak loins. A dipped or 'sway' back is a conformation weakness, whilst a 'roach' back, where the shape of the spine makes the loins convex, can produce an uncomfortable ride.

The working horse needs plenty of lung and heart room, so a broad, deep chest and good depth of girth is indicated, in proportion, of course, with the size of the horse. These aspects will develop, as will the musculature, with work and maturity. Well-sprung ribs also allow for space in the chest cavity.

Whilst a fit horse will not show a large belly, there must be room for the proper functioning of the digestive system, so a narrow, 'herring' gut is a fault. Finally, all these aspects of conformation must be put together to form an overall picture of an athletic, well-proportioned animal. If you do not like what you see, do not breed from it.

A horse is said to have 'good paces' when he moves with a regular, straight, rhythmic and forward going movement at all gaits. Action is the word used to describe the way he moves at any particular gait. For example, he might have a high knee action, as in the Welsh Cob, or, conversely, a daisy cutting action, as in the children's riding pony. Any irregularity of action indicates a defect due to poor conformation or injury, chronic unsoundness or lameness. Commonly found deviations from straight action are 'dishing' or 'paddling', when the horse swings the lower leg and foot outward and 'plaiting', when he brings one foot across in front of the other. These are the result of limb conformation defects, which may reduce the horse's athletic potential. The desired action for the riding horse is straight, free and not exaggerated in either direction. He must be able to pick his feet up to stay out of

Fig 7   Equine athletes, competing in the Arab Horse Society Marathon.

Fig 8   The typically high knee action of the Welsh Cob and ...

trouble, but should not have an uncomfortably high action.

In Germany, a horse with an uncertain temperament would be culled from the breeding system. In Britain, breeding decisions, such as which horses should be kept entire and which mares should go to which stallions, are taken by individuals. Over the years, there has been a tendency among hunter and riding horse breeders to put appearances first, performance and movement second and temperament (if it is considered at all), last. This situation has been exacerbated by the use of Thoroughbred stallions taken from the racing industry, where the criterion for selection is speed.

Fig 9   ... the daisy cutting action of the child's riding pony.

17

You will often hear it said that such and such a stallion has a good temperament – the implication being that most others do not. But what do we mean by 'a good temperament'? There is evidence to suggest that although some aspects of temperament, such as how the horse instinctively reacts to various stimuli, are inherited, the way in which he is trained and managed will have a considerable influence on his behaviour in the domestic environment.

It must be accepted that if we want to breed horses that are athletic, fast and strong, we are going to get lightning quick reactions, a high state of alertness and keen, spirited behaviour as part of the deal. The heavier Warmbloods are often condemned for lack of boldness. This is surely commensurate with their docile, placid natures. Nevertheless, if you want to breed a riding horse, do not choose a stallion who is known to be difficult to handle. Why risk bringing difficulties upon yourself?

The final point to consider when deciding to breed a foal is what you ultimately want to do with it. Which types of horse are suited to which equestrian activities? For a horse with whom you can have fun at your local riding club, work on the principles outlined above, in conjunction with the right size for your height and weight and you will not go far wrong. Horses in more specialised disciplines, however, need more specific attributes.

The English Thoroughbred was developed for racing and the driving force behind racehorse breeding policies

Fig 10   Thoroughbred blood is an essential ingredient in the event horse.

*Fig 11   These show-jumpers, waiting to compete, show good
bone, strength and substance.*

is the quest for speed on the flat. Race-horses which do not succeed in flat racing may go on to hurdling or steeple-chasing, or be sold for point-to-pointing, team chasing or other uses, although buying an ex-racehorse for eventing or hunting, for example, is a gamble best left to the experts.

The event horse needs a fair percentage of Thoroughbred blood to give him the courage, speed and scope to tackle the three phases of horse trials. He needs stamina and must be able to gallop on well. At three-day event level the ideal size is about 16.1 to 16.2 hh. Any smaller and he will usually be at a disadvantage

(although there are always exceptions, such as Mark Todd's double Olympic gold medal winner, Charisma, who is 15.3 hh); any bigger and he will be likely to be less agile and more prone to injury.

The show-jumper needs the muscular power to jump big fences combined with the suppleness and agility to allow him to manoeuvre effectively in confined spaces. He should have scope at the elbow to allow him to really stretch forward from the shoulder, but be short with a strong back and loins and powerful hind-quarters. The joints of the leg must absorb tremendous force when the horse is jumping, so they must be strong and

Fig 12   A nicely matched pair of Welsh Cobs, used for driving.

flexible, without any weaknesses which might lead to problems.

The dressage horse must look impressive in action, so the scope for powerful movement, and active, balanced paces is important. For many years, the heavy Germanic warmbloods were favoured, their temperament predisposing them towards obedience in training. Although powerful, these horses often lacked the elegance and lightness of horses with a higher proportion of Thoroughbred blood and there is a welcome trend now towards the lighter type of horse, which is making itself felt in German breeding policies as well as in other countries. Temperamentally, such horses may be more difficult to train, but the brilliance they can produce in the right hands makes the effort worthwhile.

Endurance riding requires a horse that is light-framed, with economical action and good ground covering ability. He must be forward going, responsive and have the will to go on for up to 100 miles a day, when fully trained. Good feet are essential. The Arab, Anglo-Arab, or part-bred Arab are ideal, while in America, the home of endurance riding, the Arab/Standardbred cross is favoured.

Carriage driving is becoming an increasingly popular sport and suitable animals are much sought after. The driving horse or pony has different requirements to the riding horse. A more upright shoulder is better adapted to drawing a vehicle and a compact, muscular build gives strength for size. Horses from the tiniest Shetland pony to the biggest Shire are used for driving, put to vehicles commensurate with their size and strength. Many breeds are suitable and some, such as the Welsh Cob, are traditionally dual-purpose riding and driving animals. Continental warmbloods were

Fig 13 Agile and athletic, the jumping pony has different attributes from ...

Fig 14 ... the leading rein show pony, where refinement, calmness and poise are required.

developed by introducing Thoroughbred blood to heavy carriage horses.

Finally, you may be interested in
breeding ponies and there are as many
breeds and types of ponies as there are
horses, all with their own uses. The nine
British native breeds all have their own
breed societies and enthusiasts. Some,
such as the Dartmoor, make ideal
children's first ponies, whilst others, such
as the Connemara, Highland and Welsh
Cob, carry adults easily. They are all
capable of performance work, within the
varying limits of their respective heights
and make excellent general-purpose
mounts.

There are also the 'children's riding
ponies', which are produced mainly for
the show ring and epitomise the pony at
its most refined, with a high proportion
of Thoroughbred blood. Between these
and the natives come show hunter ponies
and working hunter ponies, each with
their own role to play in the rich variety
of the equine world.

2 Selecting a Sire or Dam

The perfect horse is a paradoxical creature. He appears in many guises, but has never yet been born. To a racehorse owner or trainer he is the horse who can outrun all others, or win the Derby; to the show producer, the all-time champion of champions; to the competitive rider, an unbeatable Olympic gold medal winner. He is an imaginary embodiment of spirit, strength and power, yet he is gentle, willing and submissive. He never goes lame, nor suffers any of the ills which make horses the great levellers they are. Perhaps fortunately, he does not exist. If he did, the horse breeder's quest would lose its fascination, which lies in the attempt to improve upon what has gone before.

You will have your own ideas of what constitutes equine perfection and the variety of the equine species provides ample scope for choice. Whatever your preference, however, there are a few consistent principles.

1. A 'hot-blood' cross will add quality to a heavier, draught type of animal.
2. Arabian blood has historically been brought in to influence many other breeds and was responsible for the foundation of the Thoroughbred.
3. The outcome of breeding from animals of mixed blood is less predictable than that of breeding from pure-breds.
4. When selecting a sire, base your choice on his good points. Avoid a horse with the same faults as your mare, but do not look for the opposite extreme.

5. Selective breeding does work, statistically speaking. However, failures also occur and whilst careful selection will improve your chances of breeding the foal you want, there are no guarantees. Be prepared for a degree of disappointment.

We will look at each of these points in more detail, before going on to consider the merits of various breeds, and their suitability for crossing with the type of mare that the majority of individual horse owners might have as a breeding prospect.

Adding Quality

This is a favourite theme of modern horse breeding and has usually meant the use of a Thoroughbred stallion on whatever stock it was desired to improve. In this context, 'quality' can be defined as refinement of conformation, improved athletic potential and a keener temperament, with all the advantages these things imply in the riding or competition horse.

This policy of using Thoroughbred sires has been immensely successful in some respects – it is now recognised that a high percentage of Thoroughbred blood is a great asset in the high-performance horse. However, in two respects, the indiscriminate application of the theory has backfired. Firstly, the stallions concerned have not always been well chosen. Flat racing, naturally, has the primary call on top Thoroughbred sires and since

Fig 15 The Thoroughbred stallion, Domitor, by Busted out of Red Goddess, stands 16.1hh and has 9in (23cm) of bone.

the sole criterion for their selection is speed on the flat, this in itself is not necessarily a problem. However, many horses which show staying-power combined with jumping ability (the very attributes required in the performance horse) are gelded to go into National Hunt racing and are thereby lost for breeding purposes. Those stallions which remain available for competition horse breeding often do not have the right kind of proven performance ability to back them up and may lack the essential qualities of stamina and soundness. The only available guide is often the family pedigree and here again, the data must be derived from racing.

In addition, the Thoroughbred sire is notoriously fickle in reproducing his own ability in his offspring, so it is prepotency which must be sought, when you are looking for Thoroughbred performance. Seek out the stallion's offspring and ascertain what kind of stock he is producing. When assessing these offspring do not forget the influence of their dam lines.

Secondly, in Britain, the National Light Horse Breeding Society (HIS) scheme made Thoroughbred stallions of varying merits widely available for use on all types of mares. This commendable effort to improve the country's riding horse stock also, inevitably, produced many inferior animals, the mares being unsuitable for future breeding. Over-enthusiastic use of Thoroughbred stallions was brought about by inexpensive stud fees under the HIS scheme coupled with an increasing demand for quality competition horses. A further aspect of the trend was the dilution of good brood mare foundation stock. This was most evident in the Irish Draught breed, where the number of good Draught mares decreased alarmingly. Fortunately, efforts are being made to reverse this situation and the HIS scheme now also includes a register of approved non-Thoroughbred stallions.

The Arabian is also used to add 'quality' to other breeds, although, as we shall see, his influence is not always acknowledged.

THE ARABIAN CONNECTION

The role of the Arabian in modern horse breeding is subject to a strange dichotomy. The oldest established of the world's breeds, there is no doubt that the Arabian has exerted his influence, for the better, on numerous breeds throughout the world. Yet in many equestrian circles he is widely regarded as being 'not a proper horse'. Many faults and failings have been laid at his door:

'He may be pretty to look at, but he's too small to be useful for anything.'

'He can't jump.'

'He's temperamental and difficult to manage.'

These accusations may sometimes contain a grain of truth. For example, breeding purely for the Arabian's finely drawn, extravagantly showy appearance, for the show ring, has, in the past, led to the production of numbers of inferior, light-boned animals of little use for competitive work. However, the recent growth of Arab racing has helped to reverse this trend and other countries, for example the USA, have always maintained tougher Arabian stock. Polish blood lines are also renowned for their stamina and endurance.

The question of temperament is a human problem rather than an equine one. Arabians in the desert lived with their humans and were bred selectively for their intelligence as well as for their stamina, soundness, speed and courage, with conformation suited to desert conditions. Today, it is the Arabian's intelligence, making him less easy to handle than more lowly bred horses, which causes problems for handlers who do not have the patience and are not prepared to spend the time to win their horse's respect and trust.

Regardless of prejudice, the Arabian invariably adds quality when used as a cross and his greatest contribution was undoubtedly in the evolution of the modern Thoroughbred. All Thoroughbreds can be traced back through the male line to three oriental stallions imported to Britain in the late seventeenth and early eighteenth centuries – the Darley Arabian, the Godolphin Arabian and the Byerley Turk.

The Thoroughbred has long surpassed the Arabian in speed, and it must be said that Arabian conformation – streamlined and lacking heavily muscled hindquarters – does not lend itself primarily to jumping, although Arabs will jump willingly if trained to do so. However, for qualities of soundness, stamina and cour-

Fig 16   Prepotency is the ability of the sire to pass on his characteristics to his offspring. The Arabian stallion, Tarim ...

age, the purest Arabian stock provides an enduring reservoir.

PURE-BREDS AND CROSS-BREDS

If you have a pure-bred mare you may be faced with the decision of whether to send her to a stallion of the same breed or to opt for a cross. This will obviously depend upon the type of mare and what you want to do with the resultant offspring. Choosing the same breed will give you a fairly predictable outcome and knowledge of what the capabilities of the offspring are likely to be, whilst crossing one breed with another produces variable results. The offspring may resemble either the sire or the dam, or both.

If you do opt for a cross, it is advisable to choose a breed not too far removed from that of your mare. Thoroughbred and Arab stallions will 'nick' with almost anything, provided conformation and temperament are taken into account, many breeds having benefitted from Arab influence at some stage in the past. However, the outcome of crossing less well established breeds with each other is much less certain and, since the aim is improvement of the original stock, is best avoided.

Fig 17   ... and his three-year-old son, Taurean, out of a Thoroughbred mare.

If you have a Thoroughbred or Arab mare, your choice of suitable crossing breeds is wide, but if your mare is part-bred, or of non-Thoroughbred or non-Arab blood, it makes sense to choose either a Thoroughbred, Arab or Anglo-Arab stallion, or one of the mare's own breed.

OPPOSING EXTREMES

One golden rule of horse breeding is to aim for good conformation – not for extremes. It is sensible to choose a stallion who has strong good points in areas where your mare is lacking and one who

does not share the same faults as your mare. However, a stallion whose weaknesses are diametrically opposed to those of the mare should be avoided also, as using such a sire might easily result in an offspring possessing a combination of the worst features of both. Opposing faults have no tendency to iron themselves out in the next generation.

SELECTIVE BREEDING

The theory that if you breed enough of anything you will eventually get a champion has, fortunately, been largely superceded in the showing breeder's world by

27

more selective policies. This should give encouragement to the private owner breeder who, after all, can sacrifice the last degree of show ring superiority in favour of personal usefulness.

Although there are many mares who should never breed foals, most of those capable of following a useful competitive career are also capable of producing useful offspring. If your choice of sire is sound, the resulting foal should grow up to meet your expectations. In the final analysis, this is the only criterion that matters.

Breeds and Their Uses

THE THOROUGHBRED

The Thoroughbred was developed in England, as a racehorse, by crossing oriental stallions with English mares. These stallions were introduced specifically for the purpose of improving the racing ability of indigenous stock and besides the three foundation sires already mentioned, many other Arabian outcrosses were made before the breed became established.

The rapidity with which the breed spread throughout the world emphasises its importance, and other countries, notably USA and France, soon developed their own stamp of Thoroughbred horses, differing subtly from the originals. The famous 'blue grass' country of Kentucky is the centre of the world's richest Thoroughbred breeding industry.

The Thoroughbred is a superb riding horse, but his spirited temperament, his keen, onward going nature and split-second reactions make him a mount suitable only for confident, experienced riders. The qualities he brings to other breeds make him the favoured cross for everything from event horses to children's show ponies, and if you use a Thoroughbred on a good cross-bred mare of almost any breeding, you can expect the offspring to maintain many characteristics of the mare, with added refinement, sharper reactions and a more lively, forward going disposition.

THE ARABIAN

The Arabian has already been considered in detail and many of the comments which apply to the Thoroughbred also apply to the Arabian. Though somewhat less favoured as a cross today than his illustrious ancestors, he still possesses unrivalled superiority in terms of soundness, stamina and intelligence.

THE ANGLO-ARAB

The Anglo-Arab is officially recognised as a breed, even though new blood is continually being introduced by the crossing of pure-bred Arabians with Thoroughbreds. The English Anglo-Arab must have a minimum of 12.5 per cent Arab blood, whilst in France, where the breed has historically been more strongly established, the minimum required is 25 per cent. It is a common misconception that the Anglo-Arab must carry 50 per cent each of Arab and Thoroughbred blood.

The Anglo-Arab, at his best, combines the advantages of both progenitors – the speed and size of the Thoroughbred, with the soundness, stamina and intelligence of the Arab. His temperament is usually spirited, but equable and many proven competition horses have posses-

Fig 18   Malish, by the top eventing sire, Master Spiritus, out of a pure-bred Arabian mare, Despina, shows the ideal characteristics of the competition Anglo-Arab. He won the Arab Horse Society Marathon in 1988 and was runner-up in 1987.

sed Anglo-Arab breeding. The revival of Arab racing, including Anglo-Arabs, has given Anglo-Arab breeding a much needed boost in Britain. However, like the other 'hot-blood' breeds, Anglo-Arabs need skilled handling and management. Used on a cross-bred mare, the Anglo stallion should bring refinement, stamina and soundness.

WARMBLOODS

Warmblood is the term used to describe the type of competition horses bred on the continent, often prefixed by the name of the country concerned. The Warmblood was basically developed by introducing Thoroughbred blood to heavier horses – mainly the type used for carriage driving – and the different types of

Warmblood horse depend upon the type of horse found in the region of origin and the proportion of Thoroughbred introduced. The process is by no means finalised, the emphasis having shifted further towards the lighter type of horse in recent years.

The Trakehner is one of the best known Warmbloods. Large numbers of high-quality saddle and carriage horses were bred in Trakehnen in East Prussia, until the Russian thrust into Germany in 1944 forced a full-scale evacuation. A 900-mile trek westwards ensued, many horses being lost en route, due to starvation and other depredations of war. Those that survived formed the nucleus of the modern Trakehner – a breed which has proved its capacity for endurance, excels in many equestrian spheres and

29

Fig 19   This attractive Dutch Warmblood filly, Fatal Attraction, by Blanc Rivage out of the Warmblood mare Inumara, is a good example of the lighter type of Warmblood now favoured for competition work.

produces excellent riding and competition horses when used on Thoroughbred or half-bred mares.

Probably the most popular of the modern Warmbloods is the Hanoverian, a comparatively heavy type which has been refined by the further addition of Thoroughbred blood in recent years. The Hanoverian has proved an extremely successful show-jumper and dressage horse and has helped to bring Germany to the fore in these international sports as well as being widely in demand for export. The Oldenburg and the Holstein (a heavier breed) are other well-known German Warmbloods whilst Holland and Denmark also produce excellent horses of this type. Warmbloods are increasingly popular as competition horse sires and there is now a thriving Warmblood Society in Britain.

THE IRISH DRAUGHT

The Irish Draught is, as its name implies, the working horse of Ireland. For many years it has been the main source of the British competition horse, when crossed with the Thoroughbred. Ireland has the ideal climatic and soil conditions for horse breeding and the Irish Draught cross is a strong, usually large animal of considerable substance and muscular power, making a superb hunter or show-jumper. Event horses of predominantly Thoroughbred breeding also gain substance and strength from up to 25 per cent Irish Draught blood. Used on a part-bred mare the Irish Draught stallion is likely to produce a strong riding horse, up to weight for its size, not capable of great speed, but with good staying power, jumping ability and a kind temperament.

30

*Fig 20   The Irish Draught lends power,
substance and a calm temperament to show
and competition horse breeding.*

THE CLEVELAND BAY

The Cleveland Bay takes his name partly from the area of Yorkshire where the breed originated and partly from his distinctive bay colouring. Developed as a true dual-purpose working and riding horse, the Cleveland Bay is still popular for carriage driving as well as for riding, in the latter sphere serving as an excellent heavyweight hunter. To add refinement and speed, he is often crossed with the Thoroughbred and the strong bay colouring, with blue-black hooves, is usually passed on to the offspring. The Cleveland cross is a durable general-purpose mount, ideal for hunting and riding club activities.

Fig 21   The Cleveland Bay stallion Forest Foreman, owned by H M the Queen.

31

Fig 22   A team of Andalusian/Arabian cross horses setting out on the Windsor to Paris long-distance ride.

THE IBERIAN SADDLE HORSE

Whether called Andalusian (from the old Moorish name Al Andalus, and later Andalucia, for Spain), or Lusitano (from the old Latin name Lusitania, for Portugal), the pure-bred Iberian saddle horse is the one and the same breed whose influence has been as wide in the past as that of the Arabian. Although this is often overlooked nowadays, he dominated European equitation for centuries and was known as the 'horse of kings'. His short-coupled, muscular yet active physique, his proud carriage and superb natural balance, easily adapted to carrying a rider, are clearly seen in numerous period portraits of noblemen on horseback. Extremely kind, sensitive and spirited, intelligent and quick to respond to training, the Iberian horse can possibly claim the accolade of the world's supreme

saddle horse and his acclaimed descendants include the Lipizzaners of the famous Spanish Riding School of Vienna.

Iberian blood influenced many of the breeds we know today – the Lipizzaner, the English Thoroughbred, the Cleveland Bay, the Welsh Cob, the Irish Draught, the Connemara pony, the Kladruber, the Friesian, the Neapolitan, the Dutch Gelderlander, the Hanoverian, the Holstein, the Fredericksborg, the Knabstrup, the Barb and possibly (but not yet proven), the Morgan. The Thoroughbred also had a significant amount of Iberian blood in its foundation according to recent research which reveals that in the Tudor and Stuart Royal studs 'Spanish', i.e. Iberian, blood still dominated and contributed greatly to the line known as the Royal mares, forerunners, on the dams' side, to the Thoroughbred. The Spanish horse was also, of course, the mount of the sixteenth century conquerors of the New World. His influence spread across the Americas and contributed to the various South American breeds as well as the more familiar Appaloosa, Quarter Horse and Mustang.

Sadly, the merits of the Iberian breeds as riding horses were eclipsed by the rising stars of horse racing and the Thoroughbred, followed by the other competitive equestrian sports. However, the pure-bred Andalusian and Lusitano have a dedicated following and as crossing sires, show considerable prepotency. If your aim is to pursue equitation as an art, riding for personal pleasure rather than winning prizes, the Iberian horse has no equal.

THE APPALOOSA

The Appaloosa, developed in North America by the Nez Perce indians and prized for his variously spotted coat, is popular in many countries. A compact, muscular horse, he is also renowned for toughness and endurance. Cross-breds frequently retain aspects of the spotted colouring and when the cross is with quality stock, can make attractive and versatile general-purpose mounts.

THE QUARTER HORSE

The Quarter Horse is an extremely agile but heavily muscled breed, which is the main provider of the ranch horse, cow pony and western show horse. He was developed by crossing horses of Spanish origin with imported English 'blood' horses to produce a dual-purpose stock and harness horse, but took his name from the short-distance racing, over a quarter of a mile, at which he excelled. His heavy musculature permits fast acceleration from a standing start, suiting him for sprinting, but not for longer distance or endurance racing. His compact build

Fig 23   A leopard spot Appaloosa and ...

Fig 24   ... a blanket spotted pattern.

33

*Fig 25   Diablo Ima Tewlip, a pure-bred Quarter Horse and
one of Britain's most successful Western trained horses.*

and power also make him capable of the
lightning-quick turns necessary for cow
work and in western reining events. A
major attribute of the Quarter Horse is
his calm temperament and crosses may
make ideal pleasure horses or hacks.

THE AMERICAN SADDLEBRED

The American Saddlebred, as his name
implies, was developed to give a comfortable ride when it was necessary to spend
long hours in the saddle. He is a smooth-
gaited horse and the breed includes the
five-gaited horse which, in addition to

walk, trot and canter, is trained to perform at a four-beat pace at both slow and
faster speeds. Another offshoot is the
Tennessee Walking Horse. Saddlebreds
make excellent riding horses, but the
methods used to display them in the show
ring are felt to be unattractively artificial
by many horse lovers.

THE MORGAN

The Morgan is a very attractive breed,
whose main place has been in the show
ring, but who is also capable of hard
work. The breed evolved from just one
foundation stallion – a small working

horse of 14hh who took his name from his owner, Justin Morgan. The horse's actual breeding is obscure, though Welsh Cob, Arab and Thoroughbred have all been reputed to have had a part in his make-up. However, he was extremely prepotent in passing on his characteristics and the typical Morgan stance and outline is immediately recognisable today. Crosses are likely to be strong, with qualities of stamina and endurance.

THE AMERICAN STANDARDBRED

The popularity of the trotting race gave rise to the development of the American Standardbred, produced from Thoroughbred stock, though shorter in the leg and heavier in build. These sound, tough, straight moving horses are often crossed with Arabians in America to produce endurance horses.

Heavy Horses

The draught horse exists throughout the world in many forms, from the great Shire horses of Britain to the tough little Haflinger ponies (also used for riding) of Austria, to cite but two examples. The draught horse, in his pure form, can be ridden, but the larger breeds tend to be slow, placid, heavily muscled and cumbersome as mounts. However, some draught breeds have played a major role in the development of horses for other purposes, as we have already seen in the case of the Warmbloods and the Irish Draught.

The Shire and the Suffolk Punch, in particular, have often contributed to the production of good heavyweight hunters and show jumpers, the proportion of heavy horse blood usually being 25 per cent or less. The traditional heavy working horses – the Shire, Suffolk Punch,

Fig 26   Working Shires and ...

Fig 27 ... the versatile Haflinger, suitable for both riding and driving.

Fig 28 The shorter limbed heavy breeds, such as the Ardennais, are ideal for working in awkward situations like this steep forest.

Clydesdale, Percheron, Ardennais and others – do not have the speed and dash needed for competitive carriage driving and lighter breeds such as the Oldenburg, Dutch Gelderlander, Lipizzaner, Cleveland Bay and dual-purpose pony breeds, for example the Fell and Dales, are preferred.

Ponies

Technically, a pony is any equine of 14.2hh and under, although there are exceptions, such as the tiny Fallabella, which claims to be a 'small horse', i.e. having the characteristics of a horse and not of a pony. Most pony breeds have comparatively greater strength for size than horses, with a correspondingly adapted physical appearance. They are compact and rounded of build, have more bone, comparatively shorter, more muscular necks, smaller ears, better natural hardiness and adaptation for protection from the weather and thick, flowing manes and tails. Even the Thoroughbred children's riding pony, which is altogether a finer boned, longer necked, more elegant and graceful creature, is expected to retain an overall 'pony' appearance, rather than that of a hack or small riding horse. It will easily be deduced from this description that there is a fine distinction between what is a pony and what is a horse and in the eyes of show judges, the distinction is frequently a matter of subjective opinion.

THE WELSH PONY

Great Britain has the most versatile and distinctive collection of native pony breeds – nine in all. Among the best-known and most popular of these is the Welsh breed, which is divided into four sections.

The Section A, or Welsh Mountain pony, is a small, lively animal, ideal as a gymkhana pony or keen small child's pony. However, he may be too lively for a first mount. The Section B, or child's pony, is often considered the ideal children's pony, being more refined and slightly larger than the Section A, but forward going and keen. The Section C, or pony of cob type, is smaller than the

Fig 29   The Welsh Section C.

Section D, Welsh Cob, but possesses all of his larger relative's action, fire and presence.

The Welsh Cob himself is the supreme equine product of the principality, varying from the true riding horse type, with higher withers and a sloping shoulder, to the true driving cob of more upright conformation. His temperament is keen, eager and onward going and although he performs well in all equestrian disciplines, he would not be a good choice of mount for a beginner, whose confidence might be shaken by his enthusiasm. The Welsh Cob is popular world-wide, is extremely versatile and crosses well, particularly with the Thoroughbred and the Arabian, the latter having influenced the breed in earlier times. The result should be a super riding club type of horse and a second generation cross back to the Thoroughbred often produces a top-class competition horse.

THE CONNEMARA PONY

The Connemara is the native pony of Ireland and is a clean legged animal of riding type, usually grey in colour, extremely versatile in ability and suitable to be ridden by a teenager or lightweight adult.

THE HIGHLAND PONY

The Highland pony is the larger of the two Scottish natives and is suitable for both riding and driving. He is compact and powerful, with feathered legs, great stamina and a calm temperament, though not fast. In the Highlands, he is familiarly used to bring the deer stalker's kill down from the hill.

THE DALES AND THE FELL

The Dales and the Fell are two closely related breeds, both frequently black or dark in colour and with flowing manes and tails and long, silky 'feather'. The Fell is slightly smaller at full height, with compact conformation, whereas the Dales has a higher knee action and owes more to draught horse ancestry. All these larger native breeds cross well with the Thoroughbred to make excellent riding horses.

THE NEW FOREST PONY

The New Forest is a riding pony breed which, in the past, has been subject to many outside influences and the true type has been established comparatively recently. The breed does make a versatile mount and the larger ponies are capable of carrying adults.

THE DARTMOOR, THE EXMOOR AND THE SHETLAND

The remaining small breeds comprise the Dartmoor, the Exmoor and the Shetland. The Dartmoor is best described as a miniature middleweight hunter, whose excellent temperament, good riding conformation and paces and versatility make him the ideal first pony. The Exmoor is possibly the oldest of the native breeds and is an officially designated rare breed, although positive steps have been taken to preserve him. He is immensely strong for his size which sometimes makes him unsuitable as a child's mount. However, he can carry considerable weight, is extremely hardy and also makes a first-

Fig 30   The pure-bred Dartmoor pony.

Fig 31   The ever-popular Shetland pony class.

Fig 32   Size ...

Fig 33   ... is relative!

40

class small driving pony. The smallest of the British breeds is the Shetland, again an extremely versatile pony for his size. He excels in both ridden and driving events within this limitation, but is seldom used for crossing purposes, his size making this somewhat impractical and the outcome uncertain.

Two main preoccupations of owners thinking of breeding are the potential size of the offspring and what colour he is likely to turn out. Neither can ever be guaranteed. The Thoroughbred will always add height to any other breed, cross-breed or type and the Arabian frequently produces offspring larger than himself when outcrossed. At birth, foals will usually be in proportion to the size of the mare, even when the sire is larger. This is nature's effort to avoid foaling difficulties and the young horse has five or six years of growth in which to catch up with his genetic promise.

Size and colour are governed by the same genetic laws as other characteristics. Each successful embryo possesses thirty-two pairs of chromosomes – half of each pair from each parent – and contained in the chromosomes are genes. Where these are similar, or homozygous, the offspring will resemble both parents in that particular characteristic. For example, where the genes from both parents purely stipulate chestnut coat colour, the offspring will be chestnut. If the genes are dissimilar, or heterozygous, one will be dominant and the other recessive, the dominant gene producing its particular characteristic in the offspring. For example, the gene responsible for grey coat colour is known to be dominant over chestnut, so where a grey gene pairs with a chestnut gene, the offspring will be grey. However, this does not mean that all the offspring from a grey horse will necessarily be grey, as the horse may carry recessive genes for another colour, which might be reproduced when the horse is mated to another of that colour.

3   The Reproductive System

In this chapter the first topic we will cover is the anatomy of the male and female reproductive tracts. Then the hormonal changes that control the reproductive cycles of the stallion and mare and the events leading up to conception will be described. Some time will be spent discussing the methods used to ensure that the union of mare and stallion results in a normal healthy foal and finally, some attention must be given to those things that might go wrong.

The Anatomy of the Male Reproductive Organs

The reproductive tract of the male consists of two glands called the testes, which are suspended between the thighs in a bag of skin and connective tissue called the scrotum. The scrotum is divided into two compartments, one for each testis and each containing an inner lining called the tunica vaginalis which contains and protects the testis. The space between the tunica vaginalis and the covering of the testis is filled with a slippery fluid which ensures the easy movement of both testes in their individual sacs. The scrotum responds to temperature changes. Thus, when the temperature is high the scrotum enlarges, lowering the testes and allowing them to be cooled by circulating air. The reverse happens in cold weather when the scrotum contracts keeping the testes close to the warmth of the body. This method of temperature control is essential to maintain maximum fertility.

The testes are oval in shape and in the average-sized stallion are $10 \times 6 \times 5$ cm in size. They should be the same size, be regular in outline and resilient to the touch. The internal structure consists of a network of narrow tubules called the seminiferous tubules. It is within these tubules that the spermatozoa develop. Spermatogenesis is the name given to this process which basically involves the multiplication of a parent cell into millions of genetically different sperm cells. Two types of multiplication occur:

1. Mitosis, where the cell divides into two identical, unchanged halves, thus increasing the total number of sperm.
2. Meiosis, where the cell divides into two changed halves. The genetic material within the cell is rearranged and then halved, so that each resulting sperm cell contains thirty-two chromosomes – half the number found in normal horse cells. When fertilisation occurs, the sperm combines with the ovum shed from the mare, also with only thirty-two chromosomes, to form a single cell with sixty-four chromosomes.

Supporting the seminiferous tubules is a network of connective tissue, blood vessels and specialised cells called Leydig cells which produce the hormone testosterone. This hormone is responsible for the development of male characteristics,

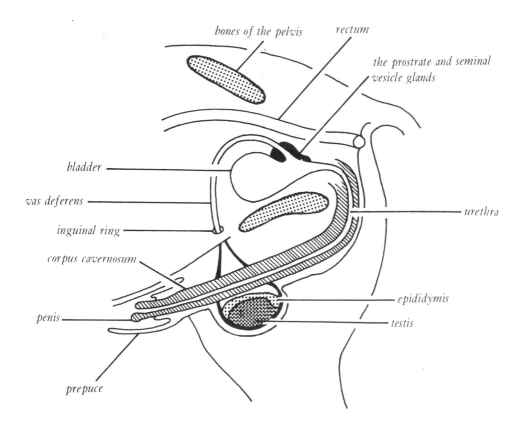

Fig 34   *The male reproductive tract.*

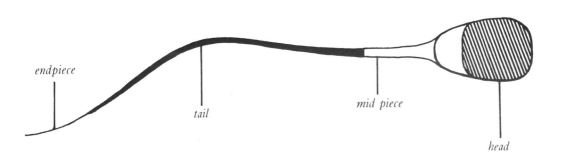

Fig 35   *A stallion spermatozoon.*

43

i.e. the heavy crest and lower pitched voice of the stallion, and it controls the correct development of the secondary sex glands.

At the front end of each testis there lies an oblong structure called the epididymis. This is a densely coiled tube which collects the sperm from the testes. As the sperm move up the tube they continue their development and by the time they reach the end of the epididymis they are fully mature fertile sperm.

From the epididymis, the sperm enter the ductus deferens. This tube is an extension of the epididymis, which allows the sperm to continue their journey to the urethra and the exterior. The ductus deferens together with the spermatic artery and vein and the cremaster muscle, make up the spermatic cord which passes through the left or right inguinal ring into the abdomen to connect each testicle to the urethra. It is through the urethra that urine is passed from the bladder to the exterior and it is in this tube that secretions from the secondary sex glands, the seminal vesicles and the prostate join with the sperm to form an ejaculate. A muscle, called the bulbospongiosus muscle, surrounds the urethra throughout the length of the penis and contractions of this muscle force the urine or ejaculate along the urethra.

The penis of the stallion is normally enclosed by the sheath (prepuce) and is only visible during urination and, of course, when the stallion is sexually aroused. It is composed of two erectile tissues, the corpus spongiosum penis and the corpus carvernosum penis. These surround the urethra and when they become engorged with blood they enlarge the penis to its erect state. When erected, the penis is a long structure with a swollen tip (the glans penis). The urethra projects past the end of the penis and is surrounded by a depression (the urethral fossa). This depression and the folds of the prepuce are often filled with a cheesy secretion called smegma.

The Anatomy of the Female Reproductive Organs

The female reproductive tract, unlike that of the male, is contained mainly within the body. It consists of two small kidney-shaped glands called the ovaries, two short tubes (the fallopian tubes), which provide a connection between the ovaries and the uterus, the uterus itself, which is closed at the caudal end by the cervix, and the vagina and vulva.

The size of the ovaries varies according to the breed and age of the horse but they are generally about 8cm long and 4cm broad. They are roughly bean-shaped and covered by a fold of the uterine ligament, a sheet of connective tissue that suspends the uterine tract from the dorsal body wall. A depression, called the ovulation fossa, is present on the concave side. Unlike other species, where ovulation occurs anywhere on the ovarian surface, the mare always ovulates through the ovulation fossa.

At a very early stage in the development of the embryo, specialised cells within the ovary multiply and differentiate into large numbers of female germ cells surrounded by primitive follicular cells. So, unlike the male, the female is born with her full complement of ova. The main function of the ovaries is to respond to the control of specialised hormones secreted by the pituitary gland

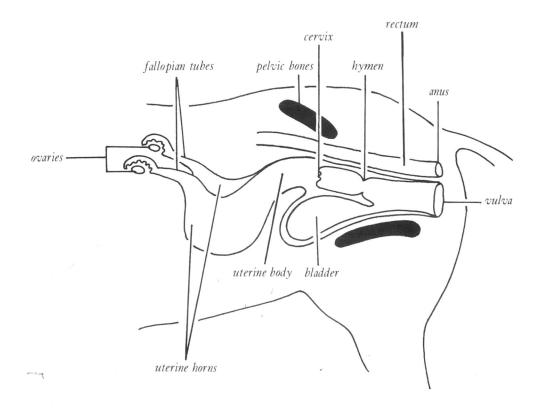

Fig 36   The female reproductive tract.

and to further develop one ova. When the ova is mature it is expelled from the ovary – a process known as ovulation. The ovary also produces hormones which help to control the process. The whole period from the development of the ova, its subsequent shedding and the development of the next ova is called the oestrous cycle.

The ovaries are connected to the rest of the reproductive tract by the fallopian tubes. One end of the tube – that closest to the ovary – is funnel-shaped (the fimbriae) and when the ova is shed through the ovarian fossa the fimbriae move over the fossa and guide the ova into the lumen of the fallopian tube where it starts its journey down the reproductive tract.

The uterus is roughly Y-shaped, consisting of a uterine body and two horns. It is in this organ that fertilisation takes place and where the fertilised ova, or embryo, starts to develop into a foal. Separating the uterus from the vagina and protecting the uterus from outside contamination, is the cervix. The muscular walls of the cervix are greatly influenced by changing hormone levels during the oestrous cycle and its appearance is of

45

Fig 37   The female reproductive tract.

great help in determining the stage of the oestrous cycle. The whole tract, from ovaries to cervix is suspended from the backbone by a strong sheet of tissue called the broad ligament of the uterus.

In the normal mare, the vagina is a collapsed tube with a muscular elastic wall. A tube, the urethra, joins the floor of the vagina a few inches internally from the vulva, this transports urine from the bladder to the exterior. The vagina provides a path from the uterus to the outside with the cervix creating a seal, internally, and the vulva, externally. A third seal, the vestibular seal, is created by a narrowing of the lumen just cranial to the urethral opening. The effectiveness of these seals is of great importance in the maintenance of optimum fertility.

The entrance to the vagina is protected by the vulva. This structure lies vertically below the anus and is composed of two upright lips, wider at the lower end.

Just within the lips, on the floor of the vulva, lies the clitoris. It is surrounded by a fossa, which, like the urethral fossa in the stallion, contains the cheesy material called smegma.

Environmental Factors Controlling Reproduction

The reproductive behaviour of both the mare and stallion is primarily governed by environmental factors. Increasing and decreasing periods of light, temperature and food supply have an important role in the development of normal reproductive activity.

In the temperate zones of the northern and southern hemispheres, the mare is a seasonally polyoestrous animal. The non-pregnant mare has a breeding year which can be divided into two. She has a period during the winter months when she shows no oestrous activity at all – the anovulatory phase – and a period, generally during the summer months, when she shows normal ovulatory oestrous cycles. These two distinct patterns of behaviour are joined by transitional periods. It is important to recognise these periods and to appreciate the changes that are occurring within the reproductive tract at this time, as they often manifest themselves as very irregular reproductive behaviour which can alarm and confuse, especially if an early conception is wanted.

In the depths of winter, the mare shows no oestrous behaviour and a manual examination will demonstrate ovaries which are small, hard and inactive. This condition is known as anoestrus. As the season advances into spring, the nights get shorter and days get longer. The mare responds to the decreasing hours of darkness by moving slowly into the ovulatory phase. Decreasing periods of darkness are not the only factors at play at this time – the concurrent increase in temperature and better food supplies also play a role, if only a subsidiary one, in the development of full ovulatory cycles. These early environmental changes have a complex and as yet not fully understood effect on the level of hormones secreted by the pituitary gland and their effect on the reproductive organs of the mare.

The three conditions – decreasing darkness, increasing food and temperature – which we can call the environmental threshold, are at their most potent some time in May or June. By then, the mare has developed a normal oestrous cycle and is at her most fertile. It is well worth remembering that the mare's natural rhythm of seasonal polyoestrus is designed to ensure that the foal is born at a time when environmental conditions are at an optimum for its survival. It is purely for our own convenience that we attempt to arrange a birth date as close as possible to the official birth date for Thoroughbreds – 1 January.

As summer changes into autumn, the increasing periods of darkness have a subtle effect on the pituitary gland which again alters the hormonal balance, and the mare sinks into a reverse transitional period leading to winter anoestrus. This period is of little importance to us, as, by that time, we hope the mare is safely in foal.

The stallion is also affected by the environmental threshold, but not to the same extent as the mare. Although he is fertile all the year around, it has been shown that his sperm count declines and his libido is less during the winter months.

The Oestrous Cycle

The oestrous cycle of the mare starts at puberty and carries on throughout the rest of her life. The cycle varies between twenty and twenty-two days in length and is separated into a period when the mare is receptive to the stallion (oestrus), which lasts four to six days, and a quiescent period, the period between each oestrus (dioestrus), which lasts sixteen days. Oestrus itself shows considerable variation, in both its intensity and duration. This is especially true at the beginning and end of the breeding season.

How do we know that a mare is in oestrus? This can be difficult, as there is a vast individual variation in the character of oestrus and the daily management of a mare can influence her oestrous behaviour. However, the average mare shows that she is in oestrus by changes in her behaviour. She will spend frequent periods crouching, as if about to urinate, but only passing small amounts of urine. She will be attracted to a stallion or gelding and will demonstrate this attraction by again crouching and assuming a position and attitude of docility. She will raise her tail to one side, contract and expand the vulva and evert the clitoris, an action called 'winking'. There is often a character change. A placid mare may become irritable and a highly strung mare, quiet and manageable. There is also often a change in the performance of the athletic mare during oestrus.

Probably the surest method of oestrus detection is using a teaser. The practicality of teasing will be dealt with in

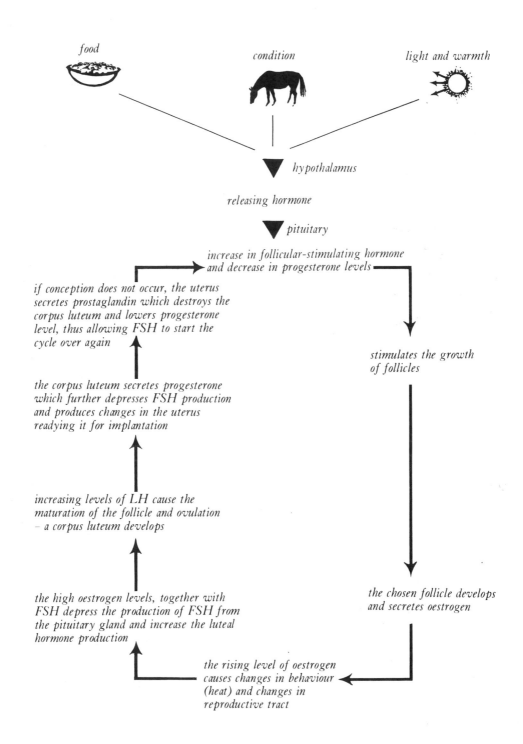

food

condition

light and warmth

hypothalamus

releasing hormone

pituitary

*increase in follicular-stimulating hormone
and decrease in progesterone levels*

*if conception does not occur, the uterus
secretes prostaglandin which destroys the
corpus luteum and lowers progesterone
level, thus allowing FSH to start the
cycle over again*

*stimulates the growth
of follicles*

*the corpus luteum secretes progesterone
which further depresses FSH production
and produces changes in the uterus
readying it for implantation*

*increasing levels of LH cause the
maturation of the follicle and ovulation
– a corpus luteum develops*

*the high oestrogen levels, together with
FSH depress the production of FSH from
the pituitary gland and increase the luteal
hormone production*

*the chosen follicle develops
and secretes oestrogen*

*the rising level of oestrogen
causes changes in behaviour
(heat) and changes in
reproductive tract*

Fig 38   A diagrammatic representation of the oestrus cycle.

Chapter 5. Enough for now to say that the way a mare reacts to the teaser, whether it be a stallion, a strange mare or a gelding, is the best way to decide if she is in oestrus.

THE CONTROL OF THE OESTROUS CYCLE

What causes the changes which occur during the oestrous cycle? Four groups of chemicals, known as hormones, secreted from the brain, the pituitary, the ovaries and the uterus, are the main orchestrators of the cycle. They act and react with each other and with the reproductive organs in the body to produce the changes which happen during the cycle. The exact method of control is not known but is likely to involve a negative feedback effect of one group on another.

Under the influence of the environmental threshold a part of the brain called the hypothalamus secretes a hormone called the releasing hormone (GnRH), which acts on a small gland tucked under the brain, called the pituitary gland, and persuades it to secrete increasing amounts of a hormone called follicle stimulating hormone (FSH) and another, called luteinizing hormone (LH). The rise in the level of these hormones is responsible for the initiation of the ovulatory season and reaches its height in the late spring.

The increasing levels of FSH circulating in the blood act on the ovaries and stimulate the growth of immature follicles. These small fluid-filled sacs each contain an egg, or ovum, and as they grow they secrete another hormone called oestrogen. Oestrogens are responsible for the behavioural signs of heat, which are necessary before the mare will accept the stallion and they also cause the physical changes which occur in the reproductive tract. The cervix, the gate between vagina and uterus, becomes relaxed, vascular and a rosy pink colour and the walls of the vagina secrete a lubricant mucus. The uterus becomes progressively larger and doughy to the feel. These changes prepare the female tract for the easy entrance of the penis and help to ensure the safe passage of the stallion's sperm as it moves up the uterus to meet the descending egg.

The increasing amounts of oestrogens, together with FSH, stimulate the follicle to produce a series of inhibitory substances which act on the pituitary gland reducing FSH production. The increasing levels of oestrogens also stimulate the luteal gland to secrete luteinizing hormone. LH, in turn, acts on the ovary and accelerates the development of one follicle, causing its eventual rupture and release of the ovum: a process called 'ovulation'. Ovulation occurs about twenty-four hours before the end of oestrus. A considerable amount of veterinary investigation is carried out in determining when ovulation occurs, as accurate timing is so important if a fertile mating is to be achieved, especially when covering a mare in hand.

What prevents more than one follicle from developing, we do not know. Sometimes, however, the regulatory factors do not work and two follicles mature and release two ova, which then develop into twins. The relatively high incidence of twinning in the Thoroughbred with its associated high level of abortion is one reason for the comparatively large number of empty Thoroughbred mares in the autumn.

Once ovulation has occurred the follicle collapses and the space formerly

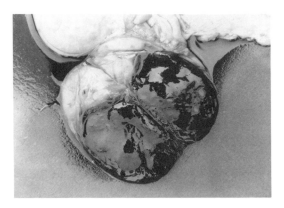

*Fig 39   The corpus haemorrhagicum,
soon to develop into the corpus luteum.*

containing follicular fluid is filled by a blood clot. The level of oestrogens produced by the follicle falls rapidly, which has the effect of terminating oestrus and the mare goes out of season. Her reaction to the stallion changes. Any advances made by him now lead to the usual behaviour of an out of season mare – biting, kicking and squealing.

The luteal phase of the cycle now begins. A mechanism known as luteinization goes into action. The blood clot, known as the corpus haemorrhagicum, begins to organise and luteal cells start to multiply, transforming the clot into the true corpus luteum. As the corpus luteum grows, it secretes increasing amounts of another hormone called progesterone. Progesterone acts on the uterus and changes it from the doughy flaccid struc-

ture that it is during oestrus into a more turgid, tubular structure. It also has a negative feedback effect on the production of LH by the pituitary gland, reducing the circulating levels of this hormone and thus slowing down the final development of the new follicle.

While this phase is going on, the amount of oestrogen is rising from its low post-ovulatory level. The gradual rise stimulates the production of a hormone called oxytocin which acts on the uterus allowing it to take a hand in the management of the cycle. If fertilisation has not taken place and implantation has not occurred, then the increasing levels of oxytocin stimulate the production of a hormone called prostaglandin on or about the fourteenth day after ovulation. This destroys the corpus luteum in the ovary and the level of progesterone produced by this gland falls rapidly. The orchestrator of the cycle – the pituitary gland – senses this drop and, in reply, increases the amount of FSH hormone, which acts on the ovary, encouraging the growth of the next crop of follicles.

It is obvious that the oestrous cycle is a complicated affair. The delicate counterplay between hormones, the dependence of the changes within the cycle on the concentration of these hormones and the effect that the environment has on the whole process mean that it is a difficult job to sort out the reason why any particular mare is a shy breeder.

4 The Stud

The prospect of sending a much loved or favourite mare to stud for the first time is worrying to many private horse owners. How will she adapt to the strange environment? What exactly is the procedure for making sure she conceives successfully? And will she be properly cared for? The term 'stud' may cover anything from a large commercial establishment with several stallions and all the up-to-date facilities, to a single stallion owner operating from small private premises.

At the top end of the bloodstock breeding market, you may find purpose-built stabling, covering sheds and exercise paddocks, with specially retained veterinary services and even private laboratory facilities. Great emphasis will be placed upon hygiene and disease prevention and mares visiting the stud just to be covered will be kept entirely separate from those in residence.

The smaller commercial establishment, standing two or three stallions, is likely to be run on sound traditional principles of horse management. There should be ample-sized foaling boxes, with an alarm system or sitting-up room where the stud groom can observe mares about to foal down without going into the box and disturbing them. There will be regular routine veterinary visits throughout the breeding season, to deal with swabbing, any fertility problems and pregnancy diagnoses.

The individual stallion owner may be involved with covering only a few mares each season and may not have the facilities to take visiting mares before foaling. Your mare might stay only for the few days when she is in season, or the owner might bring the stallion to visit her at home. This was the common practice in the days before the internal combustion engine, when the 'stallion man' would walk miles through the countryside, leading his horse and often not returning home for many weeks during the season. Nowadays, with motorised transport available, a few stallions may still be travelled to their mares, but for reasons of convenience, disease prevention and time saving, it is more practical for the mare to be brought to the stallion.

Selecting a Stud

One thing essential to the success of any stud or stallion owner is a good reputation, both for getting mares in foal, and for caring for them to their owners' satisfaction. Therefore, when you are looking for a suitable stud for your mare, word-of-mouth recommendation from someone you trust is a good starting point. You will almost certainly have chosen the stallion you want, before you think about what facilities the stud where he is standing has to offer. If possible, keep your choice of stallion open to two or three alternatives, until you have checked out the arrangements at their respective studs.

*Fig 40 Take a good look at the facilities. Are the boxes large
and airy, and everything tidy and well organised?*

The next step, before committing yourself to anything, is to visit the stud, both to see the stallion to decide if you really like him and to inspect the facilities of the stud itself. If you are not sure of your own eye for a horse and how it might combine with the attributes of your mare, take an unbiased, knowledgeable person with you. Even if you are aware of any shortcomings your mare might have, a stallion's presence can easily make him appear impressive enough to convince you that any failings he might possess are minor ones.

Visiting several studs will give you the opportunity to make comparisons, which is a valuable aid to making your final decision. Stud owners and managers will always be pleased to show you around,

but remember that they are busy people, particularly during the breeding season, so telephone for an appointment before you go.

When you arrive, take note of your first impressions – do the hedges and fences look sound and appropriately constructed for horses? Is the yard neat, clean and tidy? Are the premises well maintained? Is the general atmosphere one of calm efficiency, or is it muddled and rushed? How well do the staff handle the horses? As you are shown round, look at the facilities – are the boxes spacious and safe and are they kept clean with good deep bedding? What bedding is used? How many horses are accommodated at the stud at any given time and how much grazing is available for them? Are mares left out night and day, or brought in at night? How big are the grazing paddocks and how many mares are turned out together?

If the stud takes mares before foaling, how are these managed? Are special foaling boxes and observation facilities or an alarm system in use? Mares close to foaling should be kept separate from mares without foals. Horses are unfailingly inquisitive and a new-born foal will excite the curiosity of any other mare in the vicinity who does not have a foal of her own. Such interference before the mare and her new offspring have had a chance to get to know one another will result in an upset mare and possible damage to the foal from the intruder. The foal might even imprint the identity of 'mother' on the wrong mare. Foaling boxes should be thoroughly cleaned out and disinfected between foalings.

During the summer months a mare who is neither working, nor rearing a foal, but has come to the stud to be covered should not need hard feed or hay, provided there is adequate grazing. If grazing is limited, however, and the mare is to be kept inside part of the time, or if she is due to foal or has a foal at foot, she will need appropriate feeding and you should find out what rations the stud uses to get her used to similar food before she leaves home.

Some studs feed mares and foals in their stables, others have feeding troughs outside. Notice how the stud copes with feeding time to ensure that all animals get their fair share. Where many mares are being fed together, long hay racks with troughs beneath, such as those used for cattle, are more convenient than individual feeding bowls. However, care must be taken to see that a timid mare does not get left out.

Foals should be encouraged to eat from the first few days after birth and, given the opportunity, they will copy their dams.

Fig 41 How is feeding time organised? Are there enough feeders for everyone to get a fair share?

In the field, simply designed creep feeders can be used for foals. In the stable, they should be given their food in a separate corner from their mothers, as some mares might refuse to let their foals share the same trough. Foals should be offered the same type of food as their mothers.

On your first visit to a prospective stud for your mare, obviously the main point of interest will be the stallions standing there. When you have assessed the horse of your choice to see if he really lives up to your expectations, take a look at the way in which the stallions are managed. This can tell you a great deal about the stud and the horse himself.

Management systems for stallions tend to depend greatly on the breed concerned. Thoroughbred stallions are usually stabled, going out for exercise at some point during the day. They may be turned out into a safe paddock for an hour or two, exercised on a horse walker, or, more rarely, ridden. Mares will usually be served in hand. Native ponies, however, are much more likely to live out and run with their mares. Any combination of these two methods may be used for horses of other breeds.

The way in which a stallion is managed will have a significant effect on his mental well being as well as his physical condition. A horse who is more or less permanently stabled, except when covering mares or for brief periods of exercise, is likely to be far more keyed up, irritable and excitable than one who is running out in a natural environment or is regularly worked under saddle. If you are concerned about the horse's temperament, obviously this can be judged far more accurately in a horse who has had a ridden career than one who has never been broken to saddle.

If you are hoping to breed a competition horse and the stallion is advertised for his performance achievements, it is reasonable to expect to see him ridden and showing his ability. Most Thoroughbred stallions are seldom, if ever, ridden once they retire to stud from racing and you may have to rely entirely on racing form, blood lines and conformation. This may produce variable results, but the HIS scheme in Britain has made more Thoroughbred stallions available at relatively inexpensive fees, than other types of performance horse and many owners are happy to take up these opportunities. The importance of proven blood lines as opposed to proven performance data for any particular horse, is the subject of considerable debate.

On the continent, performance is given much greater emphasis, and prospective Warmblood stud horses are subject to a rigorous regime of testing, including performance testing, before they can be 'graded' and fully accepted as breeding sires. Thus, whilst it is fairly unusual to find performing Thoroughbred stallions, it would be equally rare to find a Warmblood stallion who was not used to being worked. Great Britain has a tradition of not riding stallions, but this is gradually changing, as the demand for sires of proven competition performance increases and a growing number of horses are imported from the continent.

In many other countries, the stallion has always been recognised as the saddle horse *par excellence*, whilst mares are retained mainly for breeding and geldings, though adequate working mounts, are far less highly prized. The Andalusian from Spain and the Lusitano from Portugal are obvious examples.

The trend towards improved perform-

Fig 42   It is easier to judge the temperament of a stallion who is regularly ridden. This Anglo-Arab, Gaillac, is trained in advanced dressage.

ance criteria has reached all levels of equestrianism and native pony stallions from all breeds can now be found, who have been broken both to ride and to drive, including the smaller breeds. Whatever your requirements, it is worth searching for a stallion who suits your purpose.

Many of the better known studs hold stallion viewing days or open days at which their horses are put through their paces for the benefit of the interested public and much can be learned by attending these occasions. Some specialise in the production of dressage horses, others in show-jumpers, or event horses and will give you an opinion on which of their stallions you should select for your particular mare and purpose.

When you make enquiries about any particular stallion, the stud will provide you with a card, giving detailed information about the horse – his breeding, going back several generations, his height and bone size and his achievements. Details of the successes of his progenitors and offspring may be noted. Finally the card should give details of the terms on which the stallion is available and his stud fee.

The Stud Fee

The stud fee is the amount you pay for the stallion to cover your mare. It does not include livery charges, feed or grass keep, fees for veterinary services, the groom's fee, nor any other incidental expenses.

The stud fee may be charged in various ways, the most simple of which is a straightforward, non-returnable fee for the covering, with no concessions. A fee on this basis is payable at the time of covering, regardless of whether the mare conceives successfully or not. Such an arrangement may mean the mare owner pays a cheaper stud fee than on other terms, but it has the obvious disadvantage that if the mare does not become pregnant, he has lost his money and has no comeback.

A common arrangement for many stallions standing at private studs is 'no foal, free return'. The fee is payable after the covering, but if the mare does not become pregnant, she may return to the same stallion the following year without a further fee being paid.

'No foal, no fee' means that if the mare does not hold a successful pregnancy, the stallion owner receives no payment for his trouble, so you can be sure the stud will do their best to get your mare in foal. With these terms, the fee becomes payable on a pre-arranged date, (usually 1 October), unless the mare is certified not in foal. A variation on this arrangement is 'no live foal, no fee'. In this case, the fee is payable after covering or at the end of the stud season, but is returned in full if the mare does not produce a live foal, who survives for at least forty-eight hours.

Finally the terms may be on a split fee basis, where a percentage is paid at the time of covering and the balance becomes due, either when the mare is tested in foal (usually 1 October), or when a live foal is born. Other, more complicated, variations in terms may be found for very valuable stallions.

Stud fees vary tremendously, depending upon the breed, blood lines, value and achievements of the stallion and will usually reflect the extent of the demand for his services. The terms of arrange-ment will depend upon the stud operator's policy, but, in general, will be comparatively more expensive, the more favourable they are to the client. Thus, 'no foal, no fee' terms would be more expensive than a straight fee, for the same stallion.

Occasionally, successful stallions are offered at an 'all in' fee, which includes livery, veterinary charges and other incidental expenses. However, you will usually have to meet these costs separately, so check in advance of booking your mare into the stud what their charges will be.

Keep charges will depend upon whether the mare is to be kept out at grass, or stabled and a higher charge will be made for mares due to foal at the stud. Veterinary costs will depend upon the extent of the work done which may include swabbing, fertility investigations and pregnancy diagnosis. These may be paid via the stud or direct to the veterinary surgeon. If the mare has to stay at stud for more than six weeks, farriery charges will also be incurred and a well-managed stud will worm all stock on a regular basis. A groom's fee is also usually payable. Livery and keep charges should realistically reflect the cost of looking after your mare and keeping her in good condition. Beware of any stud whose charges are suspiciously low.

Sending Your Mare to Stud

When you have selected a stallion and stud, book your mare in, even if you do not want to send her to stud immediately. Many good stallions are fully booked quite early in the season. The next deci-

Fig 43 Sturdy, specially designed stud fencing, with a wide avenue between paddocks.

sion is when to send the mare and this will depend primarily upon when you want the foal to be born. The mare's gestation period is approximately eleven months and foals are born any time from January to August.

Ideally, you should aim for your foal to be born between the end of April and mid-June – late enough to miss the worst weather of early spring, but early enough to make the most of the spring grass growing season. This entails having your mare covered at some point between mid-May and mid-July. Watch your mare for the first signs of oestrus and keep a record of each time she comes into season. Plan to send her to stud early enough that if she does not conceive from the first attempt, there is time to try again. You can keep on trying right through to September, if necessary. However, foals born in late July or August will need more supplementary feeding and careful management, having missed the best of the summer grass and weather.

Since the mare will be spending time in company with a number of strange horses, all of whom also come from different areas, it is sensible to make sure that her influenza vaccinations are up to date and, in case of accidents, her tetanus booster. Several days before being delivered to the stud, she should also be wormed.

The stud will insist upon the mare's hind shoes being removed, to prevent the risk of damage both to the stallion and to other mares with whom she will be turned out for exercise. She will also need a head collar in which she can be safely turned out. Leather is preferable, as it will break should it become entangled in anything, and nylon webbing will not. It should fit well and be clearly marked with the mare's name.

Insurance for the mare, whilst she is at stud, remains the owner's responsibility and will be covered by most comprehensive horse insurance policies without extra charge. However, do check your policy to be sure that your mare is covered. Cover for your mare whilst foaling and for the following month will usually require the payment of an additional premium.

Owners of brood mares who breed foals as far as possible on an annual basis often prefer their mares to foal down at stud. This has the advantage of making the most of specialised facilities and expertise and also creates the opportunity for the mare to be covered when she first comes into season a week to ten days after foaling – the period known as her 'foaling heat'. Sending the mare to stud before she foals also avoids the disruption of travelling to a new environment in a young foal's life.

The reason for covering a mare at foal heat is to avoid an accumulation of lost time in the breeding cycle, which would result, over a few years, in foals being born progressively later in the year. This is only really important for Thoroughbred and show mares and covering at the foal heat by no means always results in pregnancy, especially if the mare had a difficult time foaling. In fact, comparatively few mares, particularly Thoroughbreds, breed successfully year in year out and what usually happens is that a year is missed periodically, giving the mare a break and the breeder the opportunity to start early again the following year.

Mares who are not due to foal in the current breeding season are usually sent to stud a week or so before they are expected to come into season, the aim being to minimise the mare's stay at the stud and the consequent expense to the owner. However, the mare should be given enough time to settle down in her new environment before being covered. Telephone the stud to let them know the day and time you will be arriving, so that they can make the proper arrangements to receive your mare. Ideally, the mare will remain at the stud until she is either tested in foal, or until the time for her next heat period has passed without her coming into season. The average duration of stay for mares at stud is about six weeks.

The Covering

What happens when the mare is covered? The most natural procedure, of course, is for the stallion to run with his mares, as he would in the wild. They will then all be covered when they come into season and it frequently happens that a mare who proves impossible to get in foal when covered in hand, will conceive successfully if she is allowed to run with a stallion. Pony breeds often run out, but this is less common with the larger riding horse breeds, for a number of reasons.

A stallion who has not learned to cope with a herd of mares as a youngster is in serious danger of damage, either from mares who may be naturally aggressive, or as a result of his own over-eager attempts at covering mares who are not completely ready for him. A young horse, running out with mares, will soon receive enough kicks and rebuffs to know that he cannot take liberties with them, but must bide his time and make the proper approaches. Thoroughbred or other pure-bred stallions are often considered too valuable to be risked running out with a group of mares in a field.

If a stallion is running out with mares, he is left to get on with matters in a natural way and no special equipment is used. It is up to the stud operator to be observant and to know which mares have come into season and been covered, to get them tested in foal and send them home at

*Fig 44   A compromise between covering loose and in-hand –
the mare is held on a long rein, whilst the stallion, an
experienced horse, is left free to perform his task.*

the appropriate time. In-hand covering can be a very different matter. The extent of restraining and protective equipment used depends entirely upon the policy of the stud, but the basic procedure is the same.

Many studs like to walk their stallions along fences between the paddocks where mares are grazing to see if any are in season and these may have trying boards set into the normal fencing. Mares will be 'tried' every day, from a few days before they are expected to come into season until the end of that heat period. 'Trying' can sometimes actually help 'bring on' a mare who has been hovering on the edge of coming on heat.

Various safety precautions may be taken for the actual covering, depending upon the procedure followed by the particular stud. It is usual to put on a tail bandage to keep the mare's tail out of the way. The practice of washing the mare's genital area (and the stallion's penis, after service) with a disinfectant solution is unnecessary and modern veterinary opinion indicates that it may be detrimental to the body's natural resistance to infection.

Kicking boots may be fitted on the mare's hind feet, to give further protection to the stallion should the mare lash out. If a stallion is known to be overenthusiastic with his teeth or front feet, a leather guard may be strapped over the mare's withers and shoulders. It should not be necessary to twitch a mare who is receptive to being served, but some studs insist on twitching all mares, others will twitch any mares who are inclined to kick.

Fig 45 *If the mare is inclined to kick, these thick, felt boots will prevent damage to the stallion.*

Once a mare has been tried and shows herself ready to be covered, the stallion is led quietly up to her side in his covering bridle, on a long rein. He should approach the mare's shoulder and not her rear, which she would consider a threat.

The stallion will test the mare's acquiescence by nuzzling and nibbling at her neck and withers, gradually working his way back until he is completely aroused and in a position to mount her. The mare, especially a maiden mare, needs to be kept straight and this is most easily done with a handler on each side. They must be careful to avoid being struck by the stallion's forefeet as he mounts the mare. In some cases, a third assistant will be present to hold the mare's tail out of the way once the stallion has mounted her and, if necessary, to help guide the stallion's penis into the vulva. Ejaculation usually follows quickly and is signified by the stallion 'flagging' his tail. As soon as the stallion gets down, the mare will be walked quietly until she is calm and relaxed.

Fig 46 *A covering bridle, fitted with a chifney bit, for extra control.*

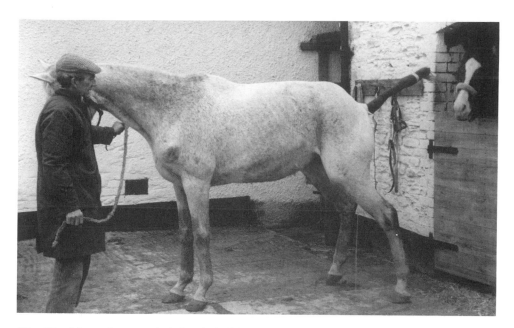

Fig 47   Here, the mare is being 'tried' by leading her up to the stallion's box. This crouching position is one sign that she is ready to be covered.

Fig 48   When the stallion mounts the mare, he may take hold of her withers in his teeth to help maintain his position. The handler, on the far side, is guiding his penis into the mare's vagina.

61

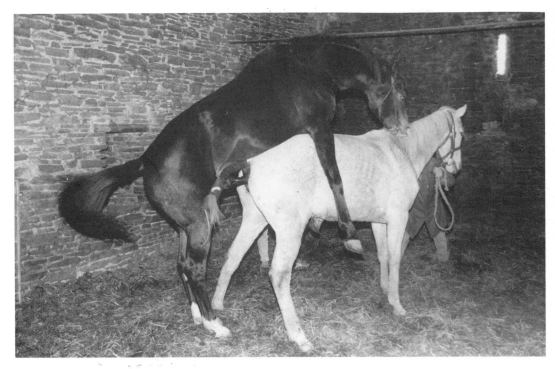

Fig 49   Once ejaculation has taken place, the stallion dismounts slowly ...

The stud will aim to see that the mare is covered at the time when she is most likely to conceive, which is around the third day of the heat period, when ovulation occurs. She will probably be covered once or twice more, as an insurance, before the heat period ends.

Fig 50   ... and is quietly led for a few minutes until he calms down. The mare will also be walked to discourage any tendency to urinate or urge, which may result in the loss of some sperm.

5 Causes of Infertility in the Horse

Managing Oestrus and When to Cover

Compared with other species, the horse is relatively infertile. The available figures vary somewhat but a conception rate of between 40 and 50 per cent to any one heat is as much as can be expected in the larger breeds of horse. Ponies tend to perform slightly better. Figures from the Hunter Improvement Society Scheme suggest that 60 per cent of mares covered by their stallions produce a foal the next year. Even when the mare is certified pregnant, the high level of abortions (15 per cent in the Thoroughbred), causes a lot of disappointment.

What can be done to improve these figures and increase our chances of a successful mating? When the stallion is running out with his mares, as in a pony herd, there is very little problem. The signs of heat are very evident, the stallion should be virile and the mare fertile. Our usual difficulty in knowing the precise time of ovulation and therefore the exact timing of service is solved as the stallion knows when the mare is at her most fertile, probably by smell or recognition of her attitude to his approaches. The problem occurs when humans interfere and cover the mare in hand.

Why do we need to cover in hand? The main reasons are as follows:

1. The need for economical use of expensive stallions. Only if the number of services per mare is carefully controlled can a successful stallion complete his book and the stud ensure that a satisfactory conception rate is maintained.
2. To make sure that each mare is actually covered and that ejaculation occurs.
3. To protect the stallion from potential injury. This fear is probably over-emphasised as observations of the reactions of mares to stallions' advances, when both are running free, suggest that the risks to either are minimal.
4. The daily stud routine of covering in hand makes it easier to check the reproductive state of the mare and to determine the best time to cover her. This especially applies to those that are shy breeders.

What is this optimum time? The reference books give a date as the last or penultimate day of oestrus. Oestrus lasts approximately four days in mid-season and longer at the beginning and end of the season. Once ovulation has occurred, the ovum remains fertilisable for the short time of four hours. After ejaculation the sperm takes about five hours to move up the reproductive tract of the female into the fallopian tubes. If the sperm and ovum do not meet during this period then fertilisation will not take place. Timing is, therefore, of the utmost importance.

How do we judge this time? We use the natural behaviour of the stallion to-

Fig 51   The expression known as 'Flehmen'.

wards a mare in oestrus and the behaviour of a mare towards other horses, towards her handlers and towards a stallion as she comes into oestrus, to judge when the time is right. This process is known as 'teasing'. During the teasing process the mare is introduced to a stallion and their reactions to each other noted.

For the protection of the stallion and the handlers the mare is generally led up to the stallion, separated from him by a board. The handlers should take care to keep sufficiently clear to avoid any chance of being kicked in the event that the mare's initial reaction is inimical. The mare and stallion should meet head to head to allow the normal introductory advances to occur. These will soon change to the typical foreplay behaviour of the stallion. He will nudge the mare's neck, often nipping her. He will display himself, neck arched and drawn up to his full height. He will often give vent to a full-blooded roar. The nudging and licking are carried on down the mare's body, along her flanks to her vulva and vagina. If the mare is in full oestrus and ready to be mated, the stallion will make it clear by showing a peculiar facial grimace known as 'flehmen'.

The mare will respond to these advances by becoming quiet and by moving towards the board. She 'shows' by raising her tail to one side and everts her clitoris. Some mares may exhibit token resistance at first, but with patient handling they become used to the situation and demonstrate oestrus. Young, maiden

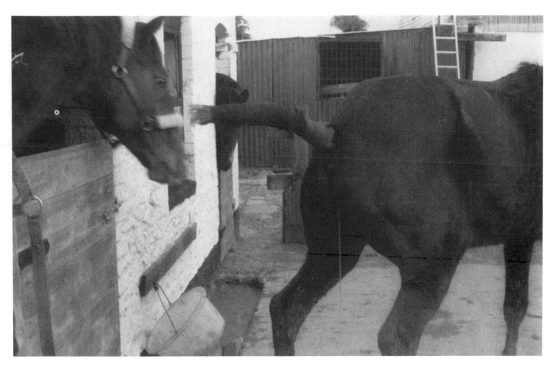

Fig 52   The typical position of a mare in a receptive condition.

mares need special care as it is all too easy to frighten them so much that mating becomes a permanently traumatic experience. Mares with young foals at foot also need sympathetic handling as their maternal instincts often override their sexual response and a long drawn-out teasing is necessary before mating instincts become stronger than maternal anxiety and oestrus can be demonstrated.

On the other hand, a mare who is in dioestrus will show her distaste in no uncertain terms. She will lay her ears back, strike out at the stallion with either fore- or hind feet and attempt to bite the handler and stallion. The more difficult mares are those that show no obvious signs either way. Are these mares actually in oestrus or not? It is very tempting to assume that the mare is in oestrus and force the issue. The situation is made even more confusing as some mares will only show when restrained in some way, generally with a twitch.

It helps to tease at the same time every day, as the familiar routine may help concentrate the mare's attitude. Mares with foals at foot may show more easily with the foal present, whilst others are happier with the foal absent. If the foal is present, an extra handler is needed to cope with the foal, especially if the teasing routine turns into the real thing and it is decided that the mare is in heat and ready to be covered. Some shy mares will not show to a stallion if led up to him, but will draw towards him if they are in an adjacent paddock. This technique works well with mares who are turned out for twenty-four hours a day. They can be teased, as a group, in the early morning and again, in the evening, when social activity is at its height. If this method is used, the fences must be suitable. It is only too easy for legs to be injured when petulant kicking thrusts them through the rails of an insubstantial fence.

To sum up, when a mare is showing well it will be easy to decide on the optimum time to cover her but this becomes much more difficult with the shy mare. Here, patient observation, recognition of subtle behaviour changes and experience are necessary if the correct timing is to be achieved. At those larger studs who have a routine which includes a daily visit from the vet, these mares can be examined internally, either manually or with a scanner and the exact day of ovulation determined. But, as a general rule, the best service intervals are two days after the onset of heat and again two days later. This method ensures the best chance of a viable ovum meeting a healthy sperm and fertilisation occurring.

Managing the Infertile Mare

The causes of infertility in the mare can be divided into two groups – those conditions non-infectious in nature which prevent a mare from conceiving and infectious causes. For the most part, non-infectious infertility is of man's doing and is the result of insisting, for economic reasons, that mares be bred at an unsuitable time of year and in a completely artificial environment. Managerial deficiencies in the handling of the stallion or mare can contribute to infertility and

physical abnormalities in either partner will, of course, have an effect. Infection, either acute or chronic, is also a common cause of infertility and has led to the development of many diagnostic procedures to demonstrate the presence of such an infection. From this knowledge we can now suggest suitable methods of control and cure.

The official birth date for Thoroughbred foals is 1 January. Therefore, in order to ensure that foals are as large and as well developed as possible when they start to race or when they enter the show ring, mares start breeding in February. Unfortunately for bloodstock breeders, a high proportion of mares are still in anoestrus or in the transition period this early in the year – the oestrous cycle is absent or irregular. Even if a normal heat is present and a successful mating is managed, ovulation and fertilisation of the ovum may not occur.

The delicate balance which exists between the emergence of normal oestrous behaviour and decreasing periods of darkness, increases in body weight and warmth (the environmental threshold), explain the erratic behaviour of the barren or maiden mare as she moves towards full reproductive activity. A sample of mares examined at this time of year will reveal a few with no ovarian activity at all – completely inactive anoestrous ones. There will be some who are beginning to respond to the environmental changes and showing signs of follicular activity but have not yet started to cycle. Others will be a stage further, showing varying degrees of cyclic activity, but still with no ovulation, and there will be a few with completely normal cycles and ovarian behaviour.

The varying changes which occur

during the transition period are best explained by imagining the mare to be hovering just under the effect of the environmental threshold. Every now and again, increasing periods of daylight, warmer weather or her improving condition, allow her to rise above this level and heat develops for a short time before disappearing as she sinks below the threshold again. Other mares remain above the threshold level and a follicle starts to develop, causing signs of heat, but because the environmental threshold is not exerting a great enough effect, the follicular development is arrested and the mare remains in heat for the full transitional period. These mares were accused of showing nymphomania but we now know that the condition is a normal physiological adjustment to environmental changes.

As the season advances from spring to early summer the all-important environmental changes now exert a strong enough effect to iron out these irregularities and a normal cycle occurs. The cycle is then repeated every eighteen to twenty-one days through the season until, in the autumn, decreasing light, colder weather and reduced food supply reverse the process and the mare moves into winter anoestrus.

Management factors also have an influence on the mare's oestrous cycle. Let us consider what happens to the hunter mare or the chaser when they are brought up to full fitness in the autumn. The increase in food, improving body condition and perhaps an extension of daylight hours, caused by artificial lighting, stimulate an active oestrous cycle at a time when this behaviour is unwanted. A decision is then made to breed from these mares, perhaps because of accident, or change in circum-

stances. Human nature being what it is, the normal routine is to rough the prospective brood mare off. The good life is now over, and the abrupt change to harsher conditions causes a lowering of the environmental threshold. This change is enough to depress the cyclic behaviour and just when normal oestrous cycles are wanted, they vanish.

How can we get a mare in foal during this confusing time? The first priority is to give the mare a thorough physical examination. We can then build up a picture of what might be going wrong and suggest ways of correcting the problem. A detailed history should be taken. The kind of information that we might ask for includes the following:

1.   Her age. The older the mare, the less fertile she becomes.
2.   Her breeding history. Does she have a foal every year or is she one of those that skips a year every now and again? Perhaps she is one of those that only breed every other year.
3.   Does she conceive on the foal heat or on subsequent ones?
4.   Have there been any problems at previous foalings? Is there a history of twinning in the family? Was there any discharge after foaling, and did the mare pass her afterbirth correctly?

An external examination should then be carried out. From this we can determine the physical normality of the external genitalia. We would check the shape of the vulva and look for any signs of previous damage or perhaps evidence of a Caslick's operation. The vulva might show signs of Herpes infection or a discharge might be present. At the same time the mammary glands could be

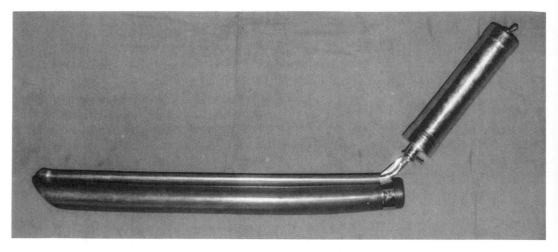

*Fig 53   A typical vaginal speculum, with its own light supply
and resterilisable plastic speculum.*

checked to make sure that they are normal.

The examination then proceeds to the internal organs. We can check these using three different methods. A manual examination allows us to palpate the cervix, the uterus and the ovaries. The cervix varies in shape and consistency, from a firm tube during pregnancy and the luteal phase of the cycle, to a soft indistinct structure during oestrus and anoestrus. The uterus is much the same, being easiest to palpate during early pregnancy and in the period between cycles. A lot of information can be gained from the shape of the ovaries. The stage of follicular growth can be estimated, and, with repeated examinations, the daily growth of the primary follicle can be monitored allowing the likely time of ovulation to be estimated.

Visual examination of the vagina and opening of the cervix can be made with a vaginal speculum. A speculum is a long tube, generally with an integral light source. When this is inserted into the vagina, the walls of the vagina and the cervix can be checked for damage, any discharge from the cervix can be sampled using a long swab, and the colour and appearance of the organs noted. If we suspect a chronic infection then the process of taking an inter-uterine swab is made much easier using a speculum.

Another method of visually examining the internal organs can be carried out with a piece of apparatus called a real time linear ultrasound. This machine creates very high frequency sound waves which are reflected by dense tissue such as bone and pass through water-filled tissues. The signals are transmitted to a VDU where the patterns can be seen. The picture can be frozen and studied, in detail, at any moment and in some machines, a photographic record can be kept.

Using this machine we can, with a little practice, 'look at' the internal reproductive organs in a great deal more detail than was previously possible. The changing

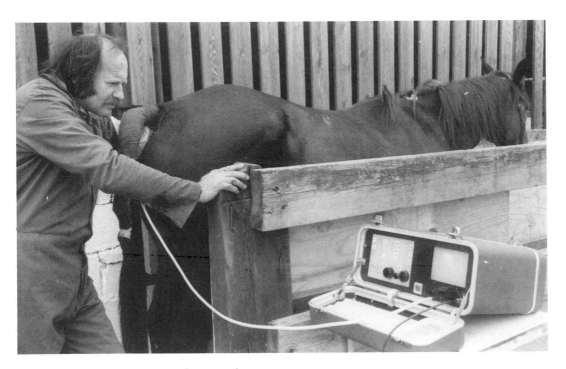

Fig 54   A realtime ultrasound scanner in use

ovary can be studied, the early pregnancy recognised and some pathological changes noted. Without doubt, our knowledge of what goes on inside the mare has increased markedly since the advent of ultrasound scanning.

One final point must be mentioned. This will be dealt with, in greater detail, in a later chapter but should be stressed now for the sake of the health and safety of the vet. To avoid broken legs and other accidents and to help in making an accurate diagnosis, the mare must be adequately restrained during these examinations.

By now, a picture of what may be wrong with your non-breeding mare should be building up, so for the rest of the chapter let us concentrate on the conditions that cause infertility and ways

of combatting them. For convenience's sake, we can divide these conditions into two main groups, those which are due to non-infectious conditions, which we will deal with first and those caused by acute or chronic infection.

NON-INFECTIOUS INFERTILITY

Anoestrus

In the depths of winter when your mare is in true anoestrus, i.e. there is no ovarian activity at all, there is little that can be done. However, if you really do want an early foal, using artificial light to increase the hours of daylight, together with extra food and warmth, will sometimes allow an earlier transition from anoestrus to ovulatory cycles. The decision to do this must be made in the early winter and the periods of extra light increased until by January the mares are exposed to sixteen hours of daylight each twenty-four hours. The best strength bulb to use is one of at least 150W, the clear type being preferable, and this should be hung above each mare 7–8 ft above floor level in a box 12 ft × 12 ft. Shadows must be avoided and the light intensity should be enough to read a page of a book with ease. Mares which start the winter off in lean but healthy condition, but then start improving under this regime, seem to respond better. Work done in this country and in USA suggests that this method will bring forward the breeding season by as much as two to three months.

The injection of the gonadotrophin releasing hormone, GnRH, one of whose functions is to stimulate the production of FSH and LH, can produce some ovarian activity. Various methods have been used, but whilst follicular growth can be stimulated, final maturation of the follicle does not occur. It appears that much more work has yet to be done before truly anoestrous mares can be persuaded to develop normal cycles.

The Transitional Period

In those mares where we have stimulated ovarian activity which falls short of full maturation of a follicle and subsequent ovulation and those mares which later in the season are going through their transition period, we can use an orally active synthetic progesterone to influence the cycle. This is administered as a food supplement called Regumate, and can be used to 'regulate' the cycle and to stimulate a normal heat and ovulation.

Regumate is used in those cases mentioned above where environmental changes, such as the increased light regime, have not had the full effect and incomplete cycles are present. Another use is to iron out the inconsistent behaviour shown in the normal transition period and to ensure that ovulation occurs. Regumate can also be used to 'manage' the heat cycle in normally cycling mares, either in the stud situation where economical use must be made of expensive stallions, or where embryo transplant techniques are in use. In the latter case, it is essential to synchronise the donor and recipient mares, to allow the fertilised ovum to be transferred from one to another. Regumate does this very efficiently.

Mares with foals at foot, as well as being shy when teased, can go for long periods with no apparent heat periods. With the present high cost of keeping a mare at stud it is essential to bring these mares on and cover them as soon as possible. The use of Regumate allows us to do this. Another use is in the competition horse. To have your mare in oestrus during an important event might not be desirable. Regumate offers a way of avoiding this by postponing oestrus.

How does Regumate work? The active ingredient contained in Regumate, a synthetic progesterone called allyltrenbolone, is added to the mare's feed, daily, for twelve to fifteen days. Allyltrenbolone has an inhibitory effect on the pituitary gland and prevents the release of LH. In turn, this prevents the normal maturation of the follicle, ovulation and development of a corpus luteum. When the treatment ends, the inhibitory effect of Regumate is removed and a bounce-back effect allows a high level of LH to be released. Theory states that a good follicle will then develop, ovulation will occur and a normal heat ensue. Unfortunately, in practice, the complexity of the hormonal control of oestrus in the mare can, in some cases, prevent the development of a normal heat after treatment.

Treatment with Regumate should prevent the development of heat, it should also block ovulation during treatment and halt the growth of the next crop of follicles. However, some mares do ovulate whilst being treated with Regumate although they rarely show physical signs of heat. In these cases ovulation is followed by a developing corpus luteum which persists until after the treatment period. The presence of this corpus luteum prolongs the dioestrous phase of the cycle and reduces the bounce-back effect which, in turn, prevents the development of the expected post-treatment heat.

Prolonged Dioestrus

Prolonged dioestrus is a condition where the period between each oestrus (normally fourteen to sixteen days), is extended for longer than normal. It is due to the presence of a persistent corpus luteum. This occurs, as we have mentioned, as a complication of Regumate treatment and can also occur in the normal breeding mare throughout the breeding season. The condition is almost impossible to diagnose by manual examination, as unlike other species the corpus luteum of the mare spends most of its allotted span deep in the substance of the ovary. However, with practice, a persistent corpus luteum can be demonstrated by the use of ultrasound imagery. High levels of the hormone progesterone in the circulating blood supply also indicate the presence of a persistent corpus luteum.

Treatment of this condition depends upon getting rid of or 'lysing' the corpus luteum. The progesterone levels then start to drop and FSH levels begin to rise, instigating the development of a normal heat. Two methods of treatment, both relying upon the same principle, are generally used to bring about the lysis of the corpus luteum.

For many years it has been known that any interference with the uterus soon resulted in signs of heat. From this observation the practice of irrigating the uterus with a saline solution, in order to stimulate oestrus, was established. It was not until the role of prostaglandins in the oestrous cycle was understood that it was realised that uterine irrigation released natural prostaglandin, which, by causing luteal lysis, instigated a normal heat.

A more practical treatment, without the attendant risk of uterine infection, is to administer synthetic prostaglandin. In exactly the same manner, the corpus luteum is destroyed allowing normal heat to develop.

Lactation Anoestrus

How often do we find that just as we want to cover a mare with foal at foot, the oestrous periods that were appearing quite normally suddenly stop? The cause is probably lactation anoestrus. This condition is caused by the hormone prolactin which stimulates the let down of milk and is present, in high levels, in the circulating blood supply of lactating mares. In some mares, these levels are high enough to block all ovarian activity and create a severe anoestrus which can be very refractory to treatment. In other, less affected, mares the high prolactin levels encourage a persistent corpus luteum, which by maintaining a high level of progesterone, prevents heat developing. In these cases, an injection of prostaglandin will destroy the corpus luteum allowing a normal heat to develop.

In between these two extremes lies a group of mares which cannot be considered to be in an anoestrous state but in which the ovaries show very little activity. The best treatment for these mares is to put them on a course of Regumate and then give an injection of prostaglandin soon after treatment.

Ovarian Tumours

The most common type of tumour affecting the ovaries is called a granulosa cell tumour. They occur in one ovary only and are mainly found in young mares. Depending on the hormone produced by the tumour, the physical signs shown by the mare can vary from complete ovarian and behavioural inactivity through to persistent oestrus. Stallion-like behaviour is a common development. Treatment consists of surgery. The offending ovary is removed and with luck, normal function is regained in the other ovary within a year.

Reproductive Tract Abnormalities

Uterine cysts   These can be present, in small numbers, in the walls of the uterus where they cause very little trouble, only becoming important when one is confused with pregnancy. Multiple cysts can develop in the endometrium (the lining of the uterus) where they interfere with fertility. There is no treatment for this type of cyst.

Cervical damage   Due to its position between uterus and vagina the cervix is often damaged, either during service or whilst foaling. Severe and repeated trauma can lead to fibrosis of the cervix and the development of adhesions which sometimes compromise the integrity of the cervical canal. Attempts to dilate the canal manually are generally unsuccessful. In these cases the ejaculate cannot pass into the uterus and infertility is the result.

Persistent hymen   Occasionally, during an internal examination of a maiden mare, a membrane can be seen at the vestibulo-vaginal junction. This membrane, the hymen, might be complete (present as a bulging wall, holding back accumulated uterine discharges), or it might partially occlude the passage. It is a relatively simple job to excise the hymen surgically and then expand the occlusion manually. Once removed, a persistent hymen should have no effect upon fertility.

Immaturity   Young Thoroughbred fillies taken out of racing, especially those that might have received treatment with anabolic steroids, can take a long time to reach sexual maturity. There is no treatment except good food and time. More than a year can pass before normal sexual function is shown.

Chromosomal abnormalities   There is a group of rare conditions which result in underdeveloped or abnormal reproductive organs. They seem to be caused by chromosomal abnormalities and result in poor reproductive performance or complete infertility. The condition may be suspected when oestrus is absent and when extremely small ovaries can be palpated per rectum. Diagnosis can be confirmed by laboratory examination of a blood sample.

INFERTILITY CAUSED BY INFECTION

Infections of the uterus constitute one of the most common causes of infertility in the mare. Two types of infection are recognised – an acute active infection which generally manifests itself after foaling and is sensitive to treatment, and a chronic type of infection which is more often resistant to treatment.

Acute Infections

Acute infections are common in the mare and are associated with the post-partum (after foaling) or post-service period. The mare is unusual among domestic animals in allowing a considerable amount of the stallion's ejaculate to enter the uterus. The foreign protein and inevitable bacteria cause a slight infection, usually confined to the endometrium. The infection is usually transient in nature and a normal, healthy mare will have cured the condition four to six days after foaling, and within hours after service.

Occasionally, a more serious infection, which involves the whole uterus, will occur. This is generally associated with a traumatic foaling or a retained placenta and if serious complications are to be avoided, must be treated immediately. The symptoms include a general malaise, a high temperature and, more seriously, an acute laminitis. Treatment consists of immediate antibiotic cover, both parentally and by uterine irrigation. Toxaemic

Fig 55 *Always check that the placenta is complete.*

shock is a common complication and attempts to control its development should be instigated immediately.

Retained placenta The mare normally expels the placenta within a few hours of birth and retention beyond this period increases the risk of an acute metritis. There are no guidelines as to the length of time which can elapse before action must be taken, but twelve hours is the accepted period before treatment is called for. After this time you should contact your veterinary surgeon who will remove the placenta manually, or, if this is not possible, by treating the mare with an intravenous injection of normal saline and a hormone called oxytocin. This causes the uterus to contract and to expel the membranes. This method has the advantage that the microscopic attachments between placenta and uterus are expelled as well, lessening the chance of an infection developing.

Acute endometritis A more common kind of infection is one which only affects the lining of the uterus – the endomet-

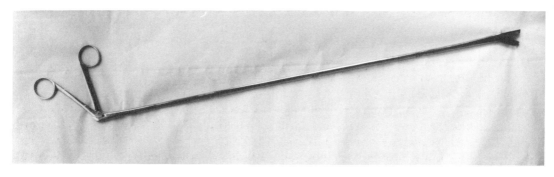

Fig 56   The type of instrument used to take a uterine biopsy.

rium. This infection is known as endo-metritis and, as already mentioned, occurs as a normal sequel to parturition and service. However, if potentially patho-genic bacteria should enter the uterus and invade the endometrium, a more serious infection takes hold. This so damages the endometrium that conception becomes unlikely. The bacteria incriminated in this infection are known as Haemophilus equigenitalis – the organism which causes CEM (Contagious Equine Metritis) – and some strains of Streptococcus, Sta-phylococcus, Klebsiella and Proteus. This group of bacteria are venereal. That is, they can be transmitted from mare to mare by an infected stallion, or indeed by any contaminated instrument used between mares.

The most common symptom of an infection of this type is a persistent dis-charge from the vulva and in such cases swabs should be taken from the uterus in an attempt to identify the organism. However, an infection can be present with no external signs and it is necessary to take a small sample of the lining of the uterus (a uterine biopsy), as well as a uterine swab, in order to demonstrate the presence of endometritis.

Contagious Equine Metritis   Of all these bacteria, the one causing CEM is the most serious. It is now a notifiable disease, i.e. the Ministry of Agriculture must be informed when it is suspected, and a set procedure has been suggested to eliminate its presence.

CEM was first recognised in Ireland during the 1976 season. It is a sexually transmitted disease which can be carried by symptomless mares or stallions. The typical disease is characterised by a pro-fuse vaginal discharge occurring twenty-four to forty-eight hours after service. This lessens over the next ten days, but inflammation of the cervix remains and can cause infertility. Antibiotic treat-ment, both parental and local, is needed to cure the condition. Early diagnosis and treatment mean early cure and a quick return to normal fertility.

In view of the serious economical risk of this disease, a series of tests have been developed that should be applied to all breeding mares. The type and scope of tests taken depends upon whether the mare or stallion could have been exposed to the disease. Details of what constitutes a high- or low-risk animal might change from year to year and up-to-date infor-mation can be obtained from the

Thoroughbred Breeders' Association, Stanstead House, The Avenue, Newmarket, Suffolk.

At the very least, visiting mares should have one negative clitoral swab before covering. Stallions should be checked before the covering season by examining washings taken from the prepuce and from samples from the urethral fossa. Routine checking of mares and stallions in this way has reduced the incidence of this disease markedly.

Coital exanthema (horse pox) Coital exanthema is an acute disease which affects the external genitalia of mares and stallions. It is also sexually transmitted and is caused by a member of the herpes virus family. Numerous vesicles develop on the external genitalia – the prepuce and penis of the stallion and the vulval lips of the mare. Secondary infection leads to acute inflammation and a vaginal discharge. The disease does not appear to cause infertility, although soreness might prevent covering and therefore cause problems in a busy stud. The infected individual can be treated with local applications of an antibiotic or antiseptic ointment and most cases heal, with no complications, within ten to fourteen days.

Other Organisms

The rest of this group of organisms cause much the same symptoms and if their presence is found by swabbing, treatment should be instigated and service delayed until a negative swab is obtained.

Treatment of Acute Endometritis

Treatment of acute infectious endometritis consists of irrigating the uterus with an antibiotic, to which the organism is sensitive, for a period of seven to ten days. The use of a plastic indwelling uterine infuser is a very convenient method of delivering the daily dose of antibiotic. The springy coiled arms, looking very like 'rams' horns' are straightened out inside an applicator, then introduced into the lumen of the uterus. Once free of the applicator, the arms spring back to their coiled position and hold the device within the uterus. The applicator can then be withdrawn

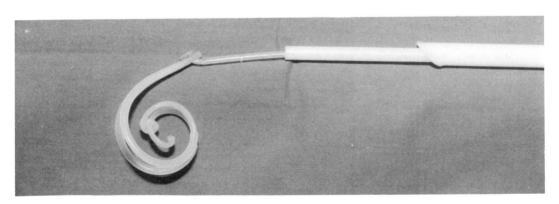

Fig 57   A ram's horn catheter. This indwelling uterine catheter makes irrigating the uterus a much easier matter.

leaving the coiled tubes and catheter in place. The end of the flexible catheter is generally sutured to the perineal region to anchor it in place. The advantage of this method is that the treatment can be administered by the owner with very little fuss and negligible stress for the mare.

Treatment of the stallion is difficult. Some of the pathogenic strains associated with veneral endometritis are particularly resistant to most disinfectants and anti-biotics. They also make life very difficult for those investigating an outbreak of venereal disease as they often cause no clinical signs in the stallion. The only practical method of preventing cross infection in these cases is the use of artificial insemination.

In the face of an outbreak of venereally transmitted endometritis, it is essential to sterilise all objects which might transmit the organism between mares. Pathological strains of Klebsiella aerogens are particularly resistant to chemical methods of sterilisation and the traditional bucket of Dettol-type disinfectant is not adequate. The use of disposable gloves and heat-resistant reusable instruments is necessary.

Chronic Infections

Chronic endometritis occurs when the ability of the mare to get rid of the normal bacterial contamination of the uterus is, for some reason, lost. This inability to eliminate the bacteria is due, in the main, to two conditions.

Vulval aspiration (wind sucking)

Faulty conformation of the vagina and vulva can allow air to enter the vagina. The constant influx of air dries and

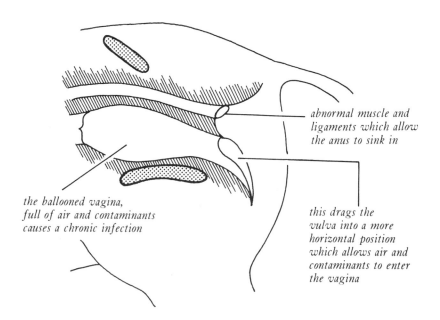

abnormal muscle and ligaments which allow the anus to sink in

the ballooned vagina, full of air and contaminants causes a chronic infection

this drags the vulva into a more horizontal position which allows air and contaminants to enter the vagina

Fig 58   The faulty conformation of a wind-sucking mare.

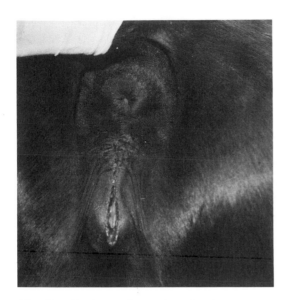

Fig 59   Faulty conformation of vulva causing the aspiration of foreign material. This vulva has been closed with a Caslick's operation.

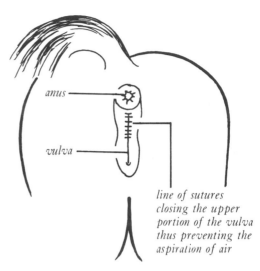

anus

vulva

line of sutures
closing the upper
portion of the vulva
thus preventing the
aspiration of air

Fig 60   The Caslick's operation.

damages the lining of the vagina and, together with the inevitable faecal contamination, it can cause a chronic infection of the vagina and cervix. This soon spreads to the uterus, where a chronic endometritis develops.

The vulval lips should remain closed during all normal movements and should be in an almost vertical position, with at least half the length of the vulva below the pelvic brim. Sometimes, because of damage during foaling, congenital malformation or old age, the shape and position of the vulva changes and the natural seal is broken allowing air to be sucked into the vagina. Mares that do this are known as 'wind suckers'.

The treatment applied in these cases is a simple operation called a Caslick's operation. This technique involves abrading the edges of the dorsal portion of the vulva and suturing them together. This closes the upper portion of the vulva and prevents the aspiration of air into the vagina. Once aspiration of air is prevented, the constant contamination of the uterus ceases and the uterus returns to normal. Where endometritis is known to be present, uterine irrigation with a suitable antibiotic may be necessary to ensure complete recovery. It is important to have your vet open the vulva with some scissors just before service or foaling in order to minimise the chance of any tearing of the vulva. If tearing should occur then the damage must be repaired at once.

Local immunity   Some mares seem to be unable to cope with the normal bacterial challenge that occurs during service because of some decrease in the level of local immunity in the uterus. The reasons for this reduced resistance are not yet known but increasing age and wear and

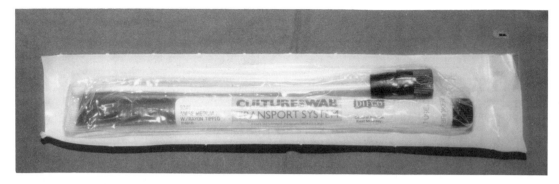

Fig 61   A suitable swab. The case contains a special
medium to ensure the bacteria remain viable during transport.

tear of the endometrium play an important role in the syndrome. This problem should be suspected if the mare develops a discharge after service, or if she returns to service before the normal nineteen days.

A diagnosis of chronic endometritis can be made from examining swabs taken from the uterus. The presence of pathogenic bacteria leads to a positive diagnosis. Special precautions must be taken

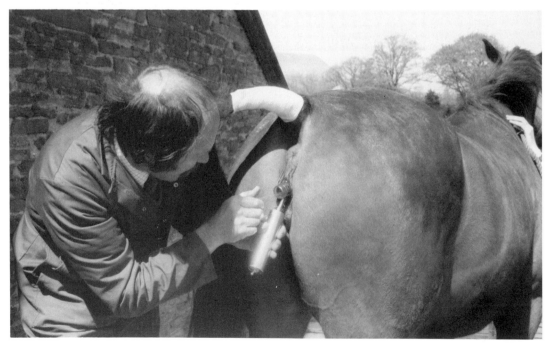

Fig 62   Taking a uterine swab.

Fig 63 Irrigating the uterus with an antibiotic solution.

to make sure the swab has not been contaminated with material from the vagina, thus confusing the issue, and it must not be assumed that a negative result means that endometritis is not present. Unfortunately, it is not always easy to demonstrate the presence of harmful bacteria and in these cases a second technique should be used. A small sample of uterine wall is taken with a special biopsy instrument and, by examining it microscopically, an assessment of the health of the endometrium can be made. Ultrasound examination of the uterus may also help in assessing the health of the endometrium.

Treatment of Chronic Endometritis

Treatment must vary according to the severity of the condition. In the uncomplicated case, uterine irrigations, with a suitable antiobiotic, are often performed. In cases where low uterine resistance is suspected, attempts to reduce the normal contamination which occurs during service should be made. The use of artificial insemination should be considered in situations where it is allowed. Unfortunately, the negative attitude of some breed societies and the Thoroughbred industry to AI restricts the use of this method of control.

Flooding the uterus with an extender

containing antibiotics before service (the minimal contamination technique), will dilute the contaminants and minimise post-service infection. Lavage of the uterus with anti-bacterial agents and saline for two to three days after service will reduce the normal post-service inflammatory response and increase the chance of normal implantation. This method can be used only in those mares who receive one service. Uterine cautery, either chemical or physical, has also been tried in cases of chronic endometritis with encouraging results.

In conclusion, we can see that the many reasons why your mare might not breed can take some unravelling. If your mare does not conceive early in the breeding season, then do not go on repeating visits to a stallion without trying to find out why she does not hold. Remember that the earlier the reasons are found out, the easier it will be to correct the problem. Also remember that although the development of the oestrous cycle, culminating in a successful ovulation, is a very complicated affair, the factors which influence it (the environmental threshold) are simple – increasing light and warmth, body condition and adequate food.

6 Management of the Pregnant Mare

The covering season begins in February and continues through to the end of the summer, so the initial care of your pregnant mare will depend partly upon the time of year when she is covered. Other factors to take into account are the amount of work she was doing before going to stud and whether you intend to continue to ride her, her physical condition and the facilities you have available.

You should make an early decision as to how you want to manage your mare during her pregnancy and then keep to it, changing her routine as little as possible. If she is fit, well and in work before covering, there is no reason why you cannot continue to ride her, if you so wish. However, strenuous exercise, such as hunting, jumping or competition work should be avoided and you may decide that it would be better to rough the mare off completely, rather than keep her in work just for quiet hacking.

There are several advantages to this, if you have sufficient pasture to turn the mare out all year round. However, exercise is essential and if you have to keep her in for a substantial part of the time, or if her paddock or field is too small for her to exercise herself adequately, steady riding out will keep her active and healthy.

A horse who has been clipped and rugged for the first half of the winter and is then covered early in the year, must be kept rugged until the weather improves in April or May. Grooming will have removed the protective grease from her coat and it will not grow thick enough to keep her warm at this time of year. Although you might not be riding her regularly, her rug should be removed and replaced each day to make sure she is comfortable and to check her condition. Light grooming, though not strapping, should be continued, to keep her skin and coat healthy under the rug, until her summer coat comes through, the weather warms up and the rug can be removed in the usual way.

Whether the mare has been sent away to stud, or covered at home, changes to her diet should be made gradually, hard feed being reduced in accordance with the reduction in her work-load. Whenever changes in routine are being made, a close eye must be kept on the mare's condition. Obviously, as she is let down from a state of working fitness, muscles will slacken – only fractionally to begin with – but it is important that good condition be maintained.

Many horses take on a lean, almost 'run up' appearance when they are exceptionally fit and if your mare is in this category, it would be advisable to decrease her work-load and allow her to put on some weight before being covered. Conversely, overweight mares should be slimmed down before going to stud, as it often proves difficult to get fat mares in foal. Once you are satisfied with your mare's condition, the aim should be to maintain it throughout her pregnancy. This will require an observant eye and a flexible attitude towards feeding.

Attitudes towards the correct management of in-foal mares vary considerably. The large commercial stud will take a different view from the smaller, private venture. The Thoroughbred breeder will have different priorities from the show horse or native pony breeder and the individual horse owner will base his decisions on personal circumstances and resources. However, two essential facilities for all would-be horse breeders are sufficient grazing and adequate stabling.

Stabling

Your stable may not be in use for much of the year, but it must be large enough – ideally, a minimum of 14×12 ft, though small horses and ponies can be accommodated in slightly smaller stabling. The size of your stable is particularly important if you intend your mare to foal down there. She will need ample room to move around, to lie comfortably stretched out and still leave adequate space for the safe birth of the foal. Once the foal is born, the stable must be big enough to accommodate both mare and foal, when necessary, giving the foal room to keep out of his mother's way. Horses do their best to avoid trampling each other, but some mares are clumsy and may inadvertently step on a foal in a confined space.

The stable should be airy and well ventilated, but not draughty. The problem of insufficient ventilation can be cured by installing air inlets at a low level, and outlets high up in the walls or roof. Draughts are most likely to occur via gaps under doors and these should be repaired or altered, as necessary, as a young foal must not be allowed to lie in a cold draught.

Attention to safety is particularly important where young horses are concerned. Of course, there should be no dangerous projections nor sharp edges in your stable, and if there are any glass windows in or near the stable they should be covered with wire mesh. An unguarded window or one protected only by bars might seem safe for months, but if a foal has to be left alone for a while, or at weaning time, he could be quite capable of jumping up and putting a foot through the glass. Never hang a hay net where an inquisitive foal might tangle his feet in it. Instead, feed hay on the floor – surprisingly little is wasted.

A straw bed is ideal for mares and foals. It should provide a thick, warm buffer from the floor and be laid sufficiently deep to prevent horses easily pawing through it to the ground. The sides must be well banked up for the foal's safety and comfort. If straw is used, it must be of good quality and never dusty. If you can obtain well-harvested oat straw, this is ideal, and eating some of it will do your mare and foal no harm at all. However, good straw is often difficult to find and poor quality, dusty straw is potentially dangerous, leading to respiratory problems. Barley straw may contain sharp awns from the ears of corn and most straw nowadays is cut short by combine harvesting, making it less easy to handle than old-fashioned long straw.

If good quality straw is not available, an alternative form of bedding is advisable and the most convenient of these is wood shavings. These are available in various grades from pre-packaged, dust-free bales, to those you bag up yourself at the local timber yard. Again, the bed should be thickly laid and banked up at the sides. If you obtain anything other than proprietary pre-packed shavings, take extra care to ensure that there are no nails or other dangerous objects mixed in with them. Many owners are reluctant to use wood shavings because of the difficulty of disposing of the soiled bedding. In fact, shavings rot down better than is usually supposed if the muck heap is properly managed. If no alternative means of disposal can be found, they can always be burned and the ash used for the garden.

Shredded paper is often used for horses who need a completely dust-free environment, but it does quickly become unpleasant, unless all the wet and soiled paper is removed every day, to prevent the accumulation of ammonia fumes. Correctly managed, it is hygienic, but an appreciable amount of paper is required to make a really thick bed and it is probably not the easiest bedding to use for mares and foals. Good-quality moss peat makes a thick, warm bed, if available, but again must be carefully managed, all the soiled and wet peat being removed each day and the remainder thrown against the walls to air.

The amount of time your mare will have to spend in her stable will be determined by the extent and quality of your available pasture. Ideally, she will live out night and day, the stable only being used when she is handled, or in emergencies, or when the weather is really bad. Decisions will have to be taken as to whether her grazing needs to be restricted if she becomes overweight in the early summer when the new grass is growing, or whether she should be brought in rather than left standing in a small, muddy paddock in winter. Remember that if your mare can only be turned out for short periods, additional exercise will be necessary.

Pasture

Horses do not need lush pasture, and breeding horses are no exception. However, there should be a good mix of nutritious grasses and herbs, the minimum of weeds and the pasture should be kept clean and well maintained. If you can arrange for the pasture to be grazed in rotation with sheep or cattle, this is the easiest way of ensuring that the land is evenly grazed and the worm burden minimised. Otherwise, the pasture should be rested periodically and the weeds cut or sprayed at the appropriate times. After spraying wait at least the time recommended by the manufacturer before restocking the land. However limited your pasture, it is better to divide it and graze each half in turn rather than utilising the entire area the whole time. Picking up droppings is time consuming, but will also help to keep down the worm burden, especially where land is heavily stocked with horses.

Harrowing, re-seeding and rolling should be carried out as necessary and fertilisation is essential to maintain healthy grass growth and cover. Chemical compound fertilisers, proprietary organic fertilisers or farmyard manure (but not horse manure) may all be used effectively. Ideally, a soil analysis should be obtained to discover the correct ratios of soil nutrients required, but otherwise follow the manufacturer's recommendations. A high nitrogen compound of say

Fig 64 Pasture stocked heavily and solely with horses will become infested with weeds and sparse in grass growth, unless proper attention is paid to management.

20:10:10 N:P:K is commonly used, but in this case, it should be spread more thinly than for farm livestock, as fast growing, nitrogen-rich grass is not required.

Organic fertilisers are increasingly popular with horse breeders, on the grounds that they are more 'natural', whilst farmyard manure is the cheapest organic fertiliser available. It should, however, be well rotted before spreading. Except on limestone soils, where it is unnecessary, occasional liming helps to maintain the correct pH (acid/alkali) value of the soil, for good grass growth. Horses must not be turned on to newly fertilised pasture until all the chemicals have been thoroughly washed into the ground, to avoid the risk of colic or poisoning. The right quality of grazing will do more than anything else to ensure the health and well-being of your in-foal mare, so the importance of good pasture management cannot be too strongly emphasised.

Mares at grass need shelter from wind and rain, but natural shelter will always be preferred to a man-made field shelter, even in the severest weather. If you have more than one or two mares together, feeding them in the confined space of a shed in order to encourage its use, might lead to disputes, quarrelling over the food supply and kicking. Trouble taken to improve the natural shelter in a field is far more rewarding. This might entail the incorporation of a windbreak in a hedge line, or planting more trees or hedges. Taking the trouble to notice where the mares like to stand to rest and feeding them near that spot is another example of thoughtful management.

Water supplies must be constant, fresh, clean and regularly inspected, to ensure that they have not become blocked up or fouled. Hedges, fences and gates must be secure and safe, particularly when foals or youngstock are present. The young equine is lively, full of explosive energy, inquisitive and lacking in fear or knowledge of danger. He has to learn about dangerous objects from his mother and his human handlers and has sometimes been aptly referred to as 'an accident waiting to happen'. Wire is a particular danger and an all too common cause of injuries which can easily prove fatal. It makes sense to take precautions and reduce risks as far as possible.

Horses are herd animals and, particularly if your mare is not being ridden, she will appreciate company. This could be other mares, or geldings. Horses turned out together in a field take a little while to establish a social hierarchy, but then usually settle down quietly. Watch to check that your mare is not bullied or kicked and that she gets her fair share of food. When feeding horses together in a field, take particular care with pregnant mares to see that the feed bowls or troughs are placed sufficiently far apart for each mare to eat in peace. If one horse gobbles her feed whilst another is a slow eater, it might be better to bring them in for feeding. Alternatively, you will have to stay until the slower mare has finished her ration, to be sure that she is not driven away from it by others.

While the grass is growing in spring and early summer, hay should not be necessary, unless you are short of pasture. At other times, it can be fed on the ground in the field, again placed in piles far enough apart for each horse to eat her share unmolested and preferably with an additional pile, so that there is always enough to go round, if one horse tries to drive another off her share.

If possible, avoid turning ponies out with bigger horses as ponies are more likely to instigate kicking matches and can easily cause a larger horse serious damage. When the time for foaling draws near, the mare should not be left with geldings or with mares who have not foaled, or are barren, as they are likely to interfere with the new-born foal, or even try to steal it, before the dam has time to recover from the birth and regain her feet. In any case, as her time approaches, the mare is likely to go off by herself so that she can give birth undisturbed, so it is better to segregate her from others who might upset her.

Routine Health Care

It is easy to put off dealing with routine health matters when your horse is turned away, but these are as important for the breeding mare as for the working horse and should not be neglected. Annual booster vaccinations for influenza should be kept up to date, otherwise you will be faced with starting a course of injections from scratch, when the mare comes back into work the following year. In any case, it is recommended that all horses should be protected from equine influenza. Even more important is protection against tetanus, an appalling disease which is nearly always fatal. All horses

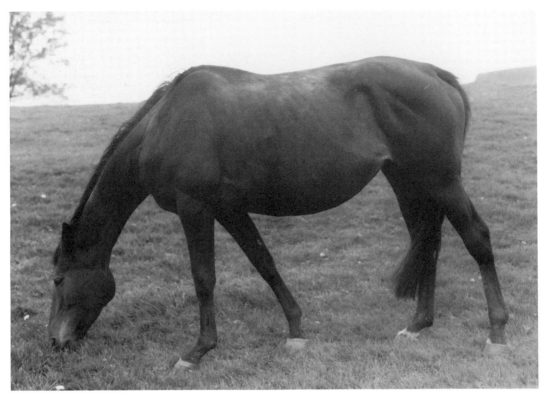

Fig 65   When foaling is imminent, the mare must be separated from geldings and barren mares. If she is to foal down indoors a suitable stable must be prepared. This mare foaled later the same night.

should be routinely vaccinated, with boosters given every two or three years, depending upon veterinary advice. Pregnant mares should receive a booster about a month before foaling, as then immunity will be passed on to the foal during the first, vulnerable months of life before he can be inoculated himself.

It is a good idea to have your mare's teeth checked and rasped if necessary, before roughing her off. Riding horses often show that their teeth need attention by refusing to accept the bit properly. Healthy teeth and gums are needed for the efficient mastication of food and your mare will need to obtain all the nutritive value available in her feed. Also, if food is

wasted, or passes through the mare's digestive system without being thoroughly digested, you will still have to pay for it. Rasping is a simple process, painless unless the situation is so bad that the mouth has become ulcerated, and will remove the sharp edges on the outside of the upper molars and inside of the lower molars which can cause trouble.

If your mare is not going to be ridden, it is preferable to remove her shoes when she is turned away. Her feet should be regularly trimmed – every five to six weeks – by your farrier, with the correct balance maintained. Neglected feet can lead to discomfort, lameness or, ultimately, to such problems as laminitis. As

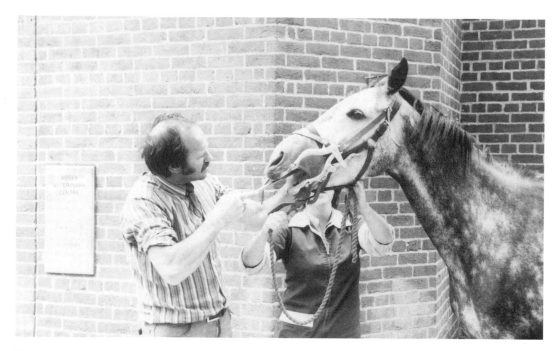

Fig 66 *Teeth need to be rasped at least once a year, to ensure proper mastication of food.*

the foal develops, the mare's weight will also increase and if her feet are overgrown or badly balanced, there is the additional risk of joint problems.

Finally, keep up your mare's regular worming programme. If she is running with other horses, all should be wormed at the same time, every six to eight weeks, but slightly more frequently in spring and early summer when the parasites are most active. Remember to change your wormer occasionally to avoid any chance of worms building up a resistance to the preparation. All horses carry worms and even if you have plenty of pasture, worming is essential.

As winter approaches, there is no need to rug your mare, even if you are riding her out occasionally. She should still be turned out for as much time as possible and although many owners think it kinder to bring mares in at night once the weather becomes colder, most horses would prefer to be left outside. The only really good reason for bringing in breeding mares at night is to save limited pasture from becoming poached and so ensure a good growth of spring grass.

A mare left out unrugged will develop a thick winter coat – much thicker than if she were rugged and in work. Rolling will also ensure that a thick layer of dust is built up in the coat, which, together with natural oils, will protect her from the weather. Grooming is not required and, in fact, should have stopped as soon as the mare was turned away in spring or summer. If she is ridden, a quick brush over to clear dried mud is all that is needed, and keep it to a minimum. This applies even to most Thoroughbred mares, who also need only be brought in

Fig 67 Two months after covering, the mare shows no noticeable signs of pregnancy.

if they are really miserable in cold, wet, windy weather. Remember that rain and wind are worse than cold, though most mares will develop protection against all but the worst winter storms.

Two problems which might occur are rainscald and mud fever. In fact, rainscald most frequently develops over the saddle area of a ridden horse, who is turned out without a rug. A horse with a thick coat, who is not being ridden, has more protection from the bacteria which cause the condition and which thrive in wet, muddy conditions. Some horses are apparently more prone to mud fever than others, and white feet are more frequently affected. Mud fever is also caused by bacteria finding a way through the pores of the skin and setting up an irritation which

develops as unsightly scabs.

If either rainscald or mud fever develop, the horse should be kept dry if possible and the scabs removed by gentle grooming. The affected areas should be treated with a mild antiseptic. In severe cases, antibiotics may be prescribed. Keep a watch on your mare in case either condition appears as they are much easier to clear up if noticed early.

Diet

What to feed is a question that worries many owners, often unnecessarily. The answer is to keep to a few basic principles and not to allow yourself to be side-tracked by the deluge of advertising for

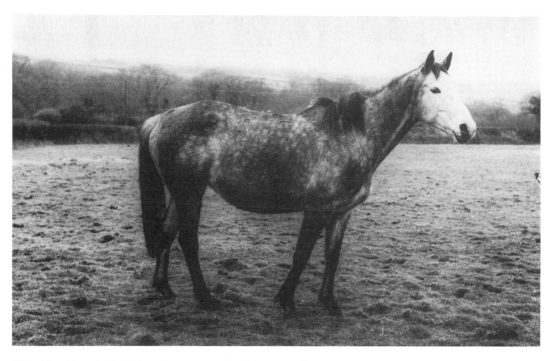

Fig 68   Roughed off in winter, around mid-term, the pregnancy is beginning to show.

all sorts of 'essential' products. Your mare needs food to repair and maintain her body tissues, to keep her warm and to give her the energy she needs to exercise herself. If she is being ridden, her food will also supply the energy for work. Once she is in foal, her food must provide the nutrients necessary for the healthy growth and development of the foal.

In the early stages of pregnancy, the embryo is tiny and it is not until the later stages that it begins to grow very rapidly. Therefore, a normal diet should be followed for the first two-thirds of pregnancy, which will keep the mare in a healthy condition. When she is out to graze on good summer pasture, no additional feed should be necessary, but from August onwards, hard feed and hay should be introduced in the usual way.

Many breeders believe that mixing rations at home gives a more consistent quality of diet than using compound mixes. Traditionally, oats, boiled barley and sugar-beet pulp are used. Oats have the benefit of being more easily digested by the horse than other grains, whilst a controlled amount of boiled, steamed or micronised barley will help put on and maintain condition. Soaked sugar-beet pulp is an extremely useful feed. It is really a 'bulk' food and makes an excellent mixer for corn and other concentrates. It is palatable, helps maintain condition and has a slightly laxative effect, which makes it particularly suitable for pregnant mares, especially those which have to be stabled and have less access to grass. It is also high in calcium, which helps offset the low calcium content of a cereal diet.

If you have the facilities for making it,

chaff is also a good mixer for hard feed, to prevent horses gobbling. It can be made from hay, or, in the old-fashioned way, from a mixture of hay and oat straw, if available. Calcium is essential for breeding stock and although the horse's feed should contain sufficient quantities, a supplement is often fed, usually in the form of limestone flour. If you wish a broad spectrum vitamin and mineral supplement can also be given, although, again, there should be sufficient vitamins and minerals for the horse's needs in good-quality food.

If you do not want to mix your own feed, manufactured compound mixes can be used instead. However, you should check the ingredients carefully. Most compound feeds also contain vitamin and mineral additives. These feeds are intended to provide a complete balanced ration and it is not advisable to combine them with other hard feed, although soaked sugar-beet pulp or chaff can be used as mixers, to prevent gobbling.

In the earlier stages of pregnancy, overfeeding is a common fault of first-time breeders, but care should be taken that the mare does not become overweight as this can lead to problems in foaling. For the last third of her pregnancy, the mare will need additional protein in her diet to cope with the needs of her growing foal. The normal protein requirement in the diet of adult horses is around 8 per cent. However, for the last third of the gestation period, the pregnant mare will require about 11 per cent protein, and foals and youngstock will need considerably more.

There are various ways of feeding the extra protein required. The simplest method for the inexperienced horse breeder is to gradually switch the mare's diet over to proprietary stud cubes, a high protein compound ration also containing vitamins and minerals. To maintain palatability, stud cubes can be fed mixed with soaked sugar-beet. The previously fed cereals or compound feed should then gradually be discontinued. Alternatively, some feed manufacturers produce high-protein balancing nuts, whilst soya bean meal, grass meal, or milk powder can also be used, added to the mare's normal ration.

Hay should be the best quality available – not hard seeds hay – containing plenty of good-quality leafy grasses. If you have only one or two mares, hay can be made available almost ad lib, once the grass has gone. If you have good hay, you can feed considerably less concentrates whilst still keeping your mare in good condition.

There can be no specific rules about how much to feed. This depends upon so many variable factors – the size, type and breed of mare, the extent of your grazing, whether she is being ridden, the quality of your feedstuffs and her own individual metabolism – that the only sure way to tell if she is getting the right amount of food is by observing her condition. If her coat is shining (even a thick winter coat should have a shine), if she is well covered with flesh but not fat, if her expression is bright and alert and if she is lively and keeps herself well exercised, you are progressing along the right lines.

If you notice condition start to fall away as the harder weather comes, increase her ration without delay. A dull coat and a disinterested attitude are signs of ill health, which should be investigated immediately. Check when she was last wormed and if you are unsure what the problem might be, call your vet.

7 Pregnancy in the Mare

Pregnancy can be defined as that period which starts when the ovum is fertilised and ends when birth of the new foal occurs. It is a period when the ovum first differentiates from its initial single cell state into two groups of cells – one group that is going to form the foetal membranes and another that will become the new individual. During this period of differentiation the latter group is known as an embryo.

The embryo starts developing from the moment of fertilisation which occurs in the mare some four to six hours after service. Fertilisation takes place in the fallopian tube and, as already mentioned, must occur quickly as following ovulation, the ovum remains fertilisable for an average of only four hours. The arrival and penetration of the sperm into the ovum stimulates the first cell divisions. These continue until, at about fourteen days post fertilisation, the mass of cells has reached a size of about 1.5cm in diameter. It is now called a blastocyst and appears as a pear-shaped sac. A group of cells, the beginning of the embryo, can be seen at the broad end. By day twenty-one the sac (by now called a yolk sac) has reached a size of 6 × 7cm and the mass of cells at the blunt end can be recognised as an embryo, surrounded by a membrane, the amnion, and connected to the yolk sac by the umbilical cord.

Over the next twenty days, further development takes place and by forty days, the yolk sac has developed into the true foetal membranes and the attach-ments between membrane and endometrium have started to form. The foetal membranes or placenta consist of two layers – the chorio-allantoic membrane which lies next to the uterus and the allanto-amnoic membrane or amnion, which surrounds the foetus. Specialised cells rapidly develop on the outside of the chorio-allantoin membrane and invade the surface of the uterus. A close contact between uterus and placenta is necessary if the large amounts of nutrients and oxygen that the growing foetus needs are to pass to it from the mare's blood supply. The umbilical cord connects the embryo to the placenta. It can be divided into two portions – an amniotic length and an allantoic portion. Within its structure lie the arteries and vein which connect the placental net of blood vessels to the developing blood system within the foetus. A tube called the uracus also passes through the umbilicus; this drains urine from the foetus into the allantoic space.

By this time, the major organs of the embryo have differentiated into their primitive selves and the embryo can now be called a foetus, which at this stage weighs about 20g and has a crown-to-rump length of 2.5cm. The eyes and eyelids, the small limb buds and the ears are recognisable and over the next twenty days the lips, nose and feet will develop. By ninety days, the shape of the hooves can be made out and the sex of the embryo can be clearly determined. At 120 days, the weight will have increased

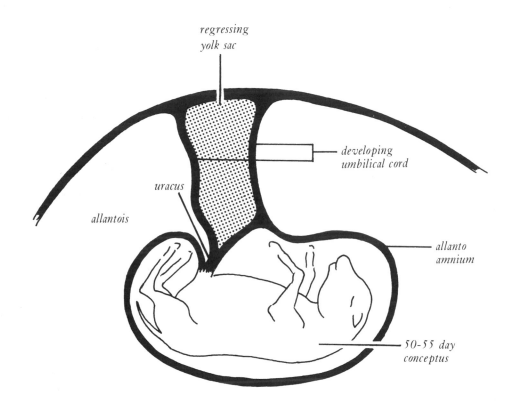

*regressing
yolk sac*

*developing
umbilical cord*

uracus

allantois

*allanto
amnium*

*50-55 day
conceptus*

Fig 69   A 50-day conceptus.

to 1kg and fine hair will start to appear on the lips, the nose and eyelids. This will have spread to the tail and mane, the back and the lower legs by day 240 and the foetus will weigh about 15kg at this stage. The covering of hair will be complete by day 320 and the physically perfect new foal will grow rapidly to its birth weight of about 45kg.

There is little change in the outside appearance of the mare during the early stages of gestation. However, as pregnancy develops, her abdomen gets larger and she becomes slower and clumsy. Some mares develop a finicky appetite or

even lose it completely as termination approaches. The udder becomes swollen and hard during the last month of pregnancy.

Internally, however, the uterus undergoes a whole series of changes as pregnancy advances. It must grow in size to accommodate the developing foetus whilst still retaining enough strength to contain the weight of foetus and foetal membranes and a perfect seal at the cervix. Somehow, it must learn to recognise the foreign placental cells which invade its internal surface. Normally, this invasion would provoke an intense reaction as the host body tries to reject the invader, but this does not happen during pregnancy.

The first signs of change occur at about fifteen days post service. The pregnant uterus becomes turgid and takes on a more narrow shape. A small swelling in the pregnant horn can be palpated at twenty-one days and by ninety days the whole uterus is filled by the conceptus and it has spread from the pelvic area forward into the abdomen. From 200 days on, the foetus can be detected per rectum and can sometimes be seen moving in the flanks when the mare is quiet.

Pregnancy Diagnosis

Why, you might ask, should I bother to find out if my mare is pregnant? Surely the fact that she has stopped horsing is enough to prove that she is pregnant. Of course, this is a good indication, but leaving it to nature can often lead to disappointment. Pregnancy is not the only reason why a mare stops coming into season and by the time you realise, in the late autumn, that your mare is empty, it is too late to do anything about it.

This method has the added disadvantage that some mares come into heat even though they are pregnant. These are the main reasons why a positive diagnosis of pregnancy is necessary. If you know, early on, that your mare is pregnant, she can be returned from stud and settled down for the gestation period. However, if the diagnosis is negative, steps can be taken with the minimum of wasted time, to find out why.

Another practical reason for a pregnancy test is to satisfy the 'no foal, no fee' requirements of some studs. The tradi-tional date when this information is needed is 1 October. Some insurance companies also require a pregnancy test before arranging cover for the pregnant mare.

There are three ways to pregnancy test a mare – manually, by rectal palpation, ultrasonic examination and hormonal tests.

MANUAL EXAMINATION

Manual examination is the traditional way to test for pregnancy. The uterus is gently palpated per rectum and, depending upon its shape and consistency, a diagnosis of pregnancy can be made. The earliest that this can be done is at about twenty-one days, but a more accurate decision can be made at the more usual time of forty-two days. As with most clinical decisions, the more experienced the practitioner, the earlier and more accurately this diagnosis can be made. As with all gynaecological examinations, the mare must be safely restrained, both for her own and the veterinary surgeon's protection.

ULTRASONIC EXAMINATION

For those wanting an early, accurate diagnosis the use of an ultra-sound scanner should be considered. This machine uses a pulse of high-frequency sound to contrast the organs of the abdomen and to show them on a screen. The sound pulses are generated from a probe which is placed in the mare's rectum and the fluid filled uterus shows up as a dark shadow with the developing embryo highlighted inside. The advantage of this system is that pregnancy can be demonstrated from

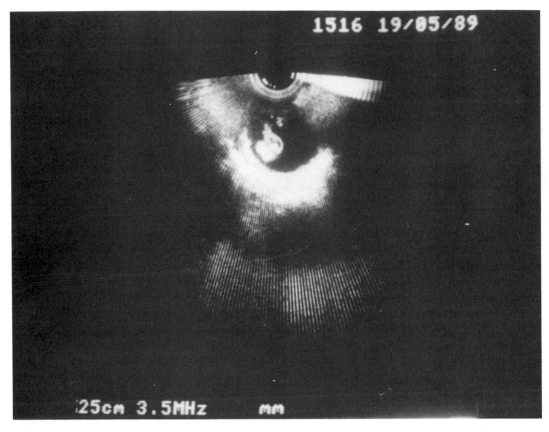

Fig 70   A photograph of an ultrasound image of a 45-day conceptus inside the dark shadow of the fluid-filled uterus.

twenty days with a high degree of accuracy and more importantly, twins can be detected at an early stage, thus allowing more time for corrective measures to be taken before it becomes too late. However it should be realised that this tool is only as good as its operator. The head must be correctly positioned and the beam of ultrasound must be pointed in the right direction to obtain results that can be interpreted correctly.

HORMONAL TESTS

These tests are performed upon samples of blood or urine. They rely on the fact that the hormone levels circulating through the body during pregnancy are different from those circulating in an empty cycling mare. Three hormones are used.

Equine chorionic gonadotrophin

This hormone is produced by specialised groups of cells present in the placenta. The hormone, known by the initials

Fig 71   Using a pregnancy testing kit in a practice laboratory.

eCG or PMSG, starts to appear in the blood by forty days post service and usually persists for a further sixty days. Blood samples taken during this period can be tested for the presence of the hormone. Kits are now available, so it should be possible to get a quick result from your veterinary surgeon.

Progesterone   This method relies upon the cyclic nature of blood progesterone levels. At eighteen to twenty days post service, just before the next heat, the amount of progesterone in the blood should be at its lowest. If the mare is pregnant, however, the persistent corpus luteum of pregnancy maintains a high level of circulating progesterone. A blood test taken at this time and showing a high level of progesterone indicates pregnancy. 'Indicates' is the operative word in this case, as, unfortunately, the test is inaccurate. High progesterone levels can be due to factors other than pregnancy, i.e. during prolonged dioestrus or following early foetal death, and due to mis-timing the sampling date.

Oestrogens   Between 150 and 300 days the placenta and the foetal gonads produce large quantities of oestrogens. A test, called the Cuboni test, can demonstrate the presence of this hormone in a sample of the mare's urine and if the test is positive, pregnancy can be assumed with a high degree of accuracy.

Failure of Pregnancy

Unfortunately a positive test for pregnancy does not always mean that a foal will be born in eleven months' time. Quite commonly in Thoroughbred mares, but less so in other breeds, pregnancy fails. If it does so during the first four months of pregnancy, the foetus is generally re-absorbed into the mare's body and little trace remains. Pregnancy failure at a later date can result in abortion, where the foetus and uterine contents are expelled, or mummification, where the foetus dies but is retained in the uterus. In these cases, the fluid portion of the contents are re-absorbed but the bones and solid tissues remain as a recognisable dried-out foal. Mummification only occurs in cases of twinning where one of the twins dies, but pregnancy is maintained with the other.

What happens to the mare after failure of pregnancy depends upon the stage of pregnancy when the failure occurs. If pregnancy failure occurs before the mare realises she is pregnant, i.e. in the five to fifteen day period, then normal cyclic behaviour is maintained and the mare returns to heat. When pregnancy fails between fifteen and forty days, the embryo is re-absorbed but the corpus luteum remains functional and the mare enters a period of prolonged dioestrus. Heat behaviour remains absent until the corpus luteum is lysed naturally or is removed by the use of prostaglandins.

Between forty and 140 days other factors influence the mare's behaviour. The influence of the corpus luteum becomes less important and the hormone, eCG, secreted by the endometrial cups (small groups of placental cells buried in the endometrium), together with the hormones progesterone and oestrogen, become more important in the maintenance of pregnancy. The endometrial cups still produce eCG after pregnancy failure and do so until they die, a period which varies considerably and over which we have no control. While eCG is present in the blood it prevents the development of normal heat cycles and therefore a mare that loses her foal during this period of pregnancy is very unlikely to breed again that season. It is these cases which cause so much disappointment, as for months the mare appears to be in foal and shows no heat behaviour, but is actually empty.

Loss of the foal after this period, from 140 days on, is a more cut-and-dried affair as once the production of eCG ceases, at any time btween 140 and 200 days, pregnancy is maintained by the foetus and placenta only. Death of the foetus then results in a rapid drop in production of foetal oestrogens and placental progesterone and these hormonal changes cause complete expulsion of the foetus and placenta. If this happens in the summer, the mare returns to heat quickly, but during the winter the mare tends to go into winter anoestrus.

What causes pregnancy failure? The reasons can be divided into two main groups, non-infectious and infectious.

NON INFECTIOUS REASONS

Twinning

Twins are a common reason for non-infective abortion. The mare's placenta is designed to support one foetus and finds it rather difficult to share this support between twins. All too often one of the pair finds that the division of resources is unequal and its share of the essential nutrients is not sufficient to support growth. At some time the smaller twin gives up the unequal struggle and dies. The other, more fortunate, twin carries on for a little time, sometimes to full term, but more often it also dies and both are aborted.

The early diagnosis of twinning is obviously important as 65 per cent of twins are aborted at around the eight month period and the mortality rate in those that survive to full term is calculated to be about 50 per cent – not a very economical proposition. Also, any manipulation has to be achieved before the endometrial cups start to produce eCG, at about forty days, as after this period an aborted mare will not breed again for several months. Luckily, the mare has her own method of re-absorbing the smaller twin so often only one embryo is left to complete its development. If twins are diagnosed then the most practical method is probably to abort both and start again.

Twisted Umbilical Cord

The umbilical cord, especially when it is too long, may twist upon itself or around the trunk or limb of the foetus. This may impede the blood circulation from placenta to foetus sufficiently to endanger the health of the foetus or even cause the death and subsequent abortion of the foal.

Congenital Abnormalities

When they are serious enough to cause the death of the developing foetus, foetal abnormalities are another cause of non-infective abortion.

INFECTIOUS REASONS

Viral Abortion

The virus Equid herpes type 1 is a virus, normally associated with respiratory symptoms, causing an upper respiratory tract infection, more common in young horses that are in contact for the first time. The virus causes a short-lived immunity and spreads rapidly through a susceptible population. When it infects a pregnant mare, it enters the body of the foetus and damages the internal organs, especially the liver and kidneys. Death of the foal follows rapidly and abortion occurs. Infection of the nearly full-term foal results in a stillbirth or the delivery of a weak, sickly foal that does not survive long. The virus can also lie latent in the body of a normal animal until activated by some stress, which, in the pregnant mare, causes damage to the foetus and abortion. Serious neurological complications can also occur in infected mares.

Control of the disease is difficult. Pregnant mares should be kept away from young animals, and, as much as possible, kept in small groups. Any aborting mares should be isolated at once and the products of the abortion should be burnt. Visiting mares should not be allowed on to infected premises nor should any mares be allowed to leave. Two vaccines against EHV1 are available in this country but only one is licenced for use in abortion protection. There is some doubt about its efficiency but from a practical point of view, frequent and regular use in the pregnant mare reduces the incidence of abortion.

97

Mycotic Abortion

The first visible sign of a mycotic abortion is generally a small, unhealthy foetus lying on the floor often surrounded by a thickened diseased placenta. The causal organism is a fungus and infection can occur by eating mouldy hay, or, more commonly, by contamination of the uterus at the time of the previous birth. Abortion due to fungal infection generally occurs later in pregnancy.

Bacterial Infection

Bacterial infection of the uterus and placenta can occur via the vagina or from the mare's bloodstream. Streptococci and coliforms are the types normally present. Local infection spreads from the vagina through the cervix and causes varying degrees of placentitis (infection of the placenta). This is common in later pregnancy when the weight of the uterus may cause vulval distortion and a developing pneumovagina with associated bacterial infection. The placentitis can be extensive enough to affect the health of the foetus and eventually cause abortion.

Septicaemic infections (those that spread from infection in the mare's bloodstream) can enter the foetus via the mare's blood supply, the placenta and the foetal blood supply. The resultant septicaemia can cause foetal death.

8 Foaling

Contrary to popular belief, the mare manages her foaling with very little problem. It is managing the owner which is difficult! The main difficulty is that only the mare knows when she is going to foal and she does so when she is ready, not when we think she ought. The gestation length is very variable as a slow-growing, small foal takes longer to reach an acceptable birth weight. By the time two weeks without much sleep have passed by, tempers have become uncertain and judgement faulty. If the delay becomes unbearable, ask your veterinary surgeon to examine the mare. He will check that all is well and will probably tell you that all you can do is wait. If you are taking some holiday to 'look after' the mare then organise it after the expected date. Your help will be better applied to mare and foal than to the mare, during foaling.

As I have said, timing the exact moment of birth is almost impossible but there are a few clues that birth is immi-

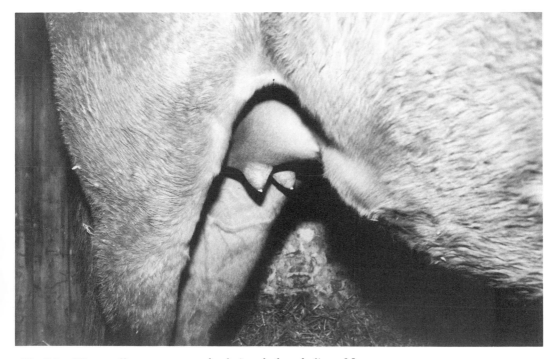

Fig 72   *The swollen mammary glands just before foaling. Note the drops of secretion that signify waxing.*

nent. The mammary glands become very tense and a honey-coloured secretion forms at the end of each teat, a phenomenon called waxing. Changes occur in the chemical consistency of the colostrum, the first milk which gives us one way of determining the onset of parturition. The levels of calcium and potassium salts increase and sodium decreases just before foaling. By measuring the concentration of these salts every day, the time of birth can be estimated to within twenty-four hours.

Other physical changes that occur are less dependable. The pelvic ligaments slacken and the musculature around the anus and vagina relaxes. The behaviour of the mare alters – first she becomes antisocial, then nervous. Unfortunately, the margin of error is wide and these signs can occur several times before foaling actually starts. One fact that has emerged from the large amounts of information collected about foalings is that the vast majority of births occur during the hours of darkness. Foaling can be induced by the administration of either prostaglandin or oxytocin – two hormones associated with birth – but the attendant risks may outweigh the advantages. There is some difficulty in judging the right time to induce as if it is done too early, the health of the foal might be jeopardised.

Labour can be divided into three stages and generally the last stage – the passing of the afterbirth – is the only stage you will see. However, if you are lucky enough to observe the whole of foaling the first signs are very much like colic. The mare shifts her weight from foot to foot, she is generally uncomfortable and outbreaks of sweating occur as she paces anxiously around the box. She passes frequent, small amounts of faeces and urine and will look around at her flanks, very often holding her tail up. This is the start of the first stage.

The signs of discomfort are due to the first waves of contraction of the uterus. These periods of contraction are brief to begin with and the mare is often relaxed enough to wander around the box, and perhaps to eat some hay, before more contractions cause another bout of discomfort. Some mares seem hardly bothered at all, others, generally the younger ones, can become very upset. False alarms often occur, especially when the mare is disturbed. The signs of discomfort disappear and the second stage does not materialise. Several of these false, first stage periods might wax and wane before the transition to the second stage occurs.

In late pregnancy, the foetus lies on its back with its legs folded to its belly but as the first stage advances, it gradually changes position. At first, the head and forelimbs flex and extend as if trying to butt the cervical opening. Indeed, this may be one of the factors that initiates cervical relaxation. Then the fore body twists into a dorsal position. As the intensity of the contractions builds up, the cervix starts to open and the uterine pressure forces the foetal membranes into the vagina. They can sometimes be seen protruding from the vulva rather like a thick white balloon.

The foetus has, by now, turned so that the head and forelegs are the right way up and are engaged in the pelvic girdle. The hindquarters are still upside-down. When the membranes rupture, releasing ten to twenty litres of amniotic fluid, the second stage of labour can be said to have started.

Noting the onset of the second stage is important as, from now on, labour is

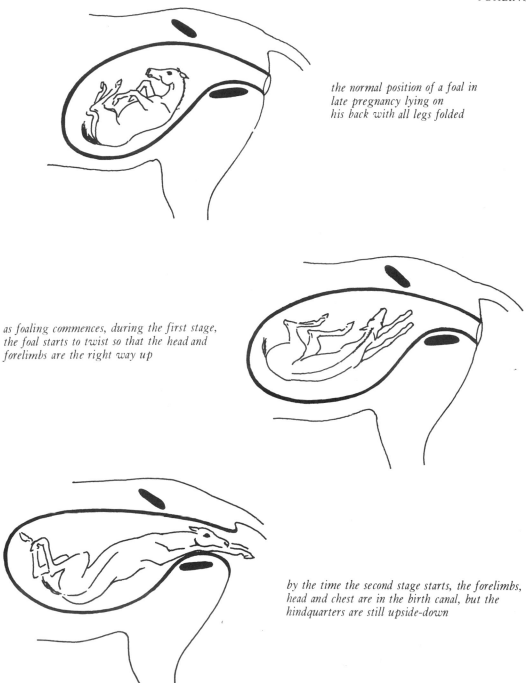

the normal position of a foal in
late pregnancy lying on
his back with all legs folded

as foaling commences, during the first stage,
the foal starts to twist so that the head and
forelimbs are the right way up

by the time the second stage starts, the forelimbs,
head and chest are in the birth canal, but the
hindquarters are still upside-down

Fig 73   Series of three diagrams showing how the foal changes
position during parturition.

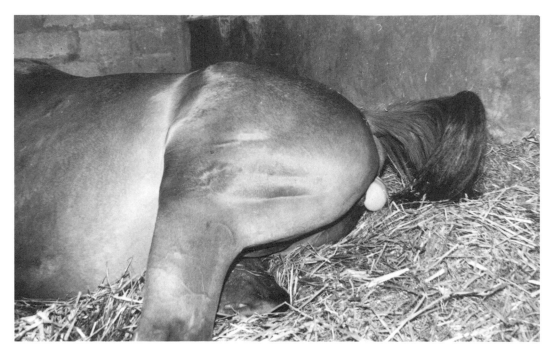

Fig 74   Foaling from start to finish.

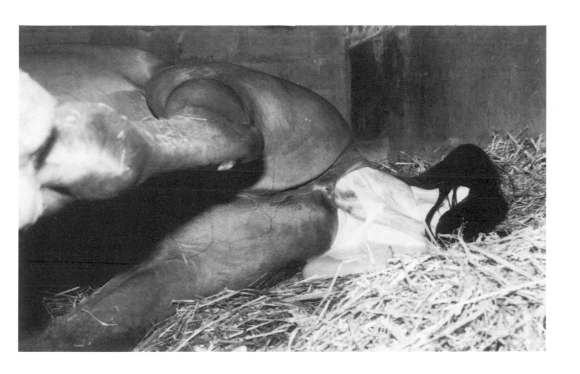

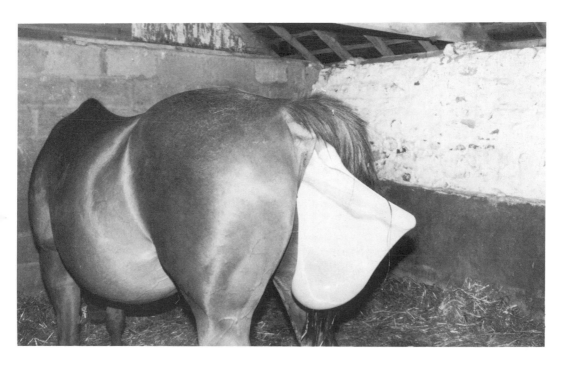

irreversible and should only take forty minutes or less to complete. The mare lies down and starts to strain, attempting to push the foal further into the pelvic girdle. At first, she will get up and down several times, (to help re-position the foetus, we think). How the mare and foetus manage to time and co-ordinate this complicated series of movements is not understood, but the frequent getting up and down of the mare, the contractions of the uterus and the foal's own movements are all involved. One rather appealing habit which the mare adopts during the second phase of birth is her tendency to lick areas where the amniotic fluid has spilt. This will include your own person if you are present. It is not known why this happens. Perhaps it is a way of marking out the 'maternity ward'. As the second phase proceeds, the mare spends more time lying down, frequently raising her head to look at her flanks.

The contractions increase in power and severity and soon the front feet can be seen, enclosed in the shiny membrane of the amnion, one just before the other, followed shortly by the nose, head and chest of the foal. It is at this stage that the foal's movements should break the membrane. If this does not happen, the amnion should be ruptured to free the foal and allow him to breath. The whole process is fast and violent, taking only five to fifteen minutes. The second phase is completed as, after a short rest, one last effort produces the foal's hips and hindquarters.

It is now that the mare should not be disturbed, as if she should get up, the umbilical cord could break and the last flow of blood from the placenta into the foal be lost. The loss of this quantity of blood could make the difference between survival and death of the new-born foal.

The hind legs often lie in the vaginal passage for a few minutes which is perhaps nature's way of plugging a potential gateway for infection to enter the uterus.

After this essential rest, which can seem to last for ever for the watching attendants but can be up to forty minutes, the mare rises to her feet to turn and inspect the new arrival and to start the cleaning process. This action imprints the image of his mother in the foal's mind and stimulates the reflexes necessary to cope with the outside world. During the resting period, another vital change takes place – the umbilical cord becomes brittle. This allows either the foal's movements or the mother rising to break the cord at the correct place (about 1.5in from the abdomen). This physical change occurs in order to minimise the chances of infection entering via the cord or haemorrhage occurring from the stump. All that needs to be done at this time is to treat the end of the cord with an anti-bacterial powder or solution. At the same time, the part of the placenta that is visible should be bundled up and tied into a parcel. Some mares, especially young ones, are liable to become alarmed at this thing that is flapping about behind them and tying it up reduces this risk.

Luckily, there is little which can go wrong during this stage, but anything giving cause for concern should be treated as an emergency. Once the strong contractions have started, the foal should arrive within ten to twenty minutes. If not, get in touch with your veterinary surgeon immediately. In all probability, he will arrive just in time to congratulate you and the mare on a strong, healthy foal but a wasted journey is much better than a dead foal or a damaged mother.

The expulsion of the afterbirth (the

placenta) takes place about an hour after the birth of the foal. During this third stage, the contractions of the uterus, necessary to expel the placenta, can cause some discomfort and the mare can show colicky signs. These are quite normal and unless they are very severe they can be ignored. The placenta is expelled inside-out, rather like pulling a sock off, and once passed it should be spread out and examined to check whether it is all there. Any bits left in the uterus could cause acute endometritis. If the mare has not passed the placenta within twelve hours, professional help will be needed to remove it. Your veterinary surgeon will probably remove it manually or treat the mare with drugs which will help her to expel it herself. Anti-bacterial drugs will also be needed to minimise the risk of infection. The appearance of the placenta should also be noticed. Any diseased, thickened bits should be kept and shown to your veterinary surgeon, since these areas can indicate a diseased uterus which might have affected the foal's intrauterine development and may possibly affect his future.

What can go wrong during this whole process? Various surveys have shown that only 4 per cent of Thoroughbred mares have trouble during foaling. This figure is probably less for other breeds, so the chances of an abnormal birth occurring with your mare are remote. However, there are various stages during birth which, if delayed, suggest that something is wrong.

1. The second stage of parturition should be a continuous process with no long periods of inactivity.
2. The fluid-filled amnion should be visible at the vulva, during contractions, soon after the start of the second stage of parturition. That is, within five to ten minutes of the start of strong continuous contractions.
3. The front feet, one often behind the other, should be followed by the tip of the nose which should be at the level of the knees.
4. The front feet, head and chest should be followed by the hips. Any delay could mean that the hips have become obstructed in the mare's pelvis.

If any problems are detected, then an examination should be made to find out what is delaying the birth process. The most common reason is a malpresentation, i.e. the foal is coming out the wrong way. These vary from a slight deviation from the norm, which if immediately recognised and corrected can allow the smooth continuation of the second stage and completion of birth, to a more serious malpresentation which will require the immediate attention of your veterinary surgeon.

How do we go about this examination? Firstly, the mare's head must be restrained by someone who can keep her as quiet as possible. Then, with hands and arm well lubricated, slowly and gently insert your hand into the vagina. Sometimes, the force of her contractions will make this quite difficult. In these cases, do not fight against her; wait until she finishes that contraction before extending your examination. We can group the types of malpresentation that can be corrected easily as follows:

1. One or both forelimbs flexed at the elbow, recognisable because the front foot or feet will be at the same level as the nose. Gentle pulling on one foot whilst

pushing the head back will generally correct this condition.

2. Crossing of both front feet. The head needs to be pushed back whilst the front feet are uncrossed and pulled forward.

3. One or both front feet bent at the fetlock. The feet are generally flexed under the cannon, but occasionally they can be flexed dorsally with the feet pointing up towards the roof of the vagina. In the former case, where the foal's foot is bent under the forearm, the malpresentation can be corrected by repelling the head so as to gain a little working room, and uncurling the flexed limb with the foot in the palm of the hand. This ensures that the sharp toe does no damage as the limb is straightened. In the latter case, the foal has often failed to rotate completely. The resultant mix-up is best left to your veterinary surgeon to sort out as it is very easy for the feet to penetrate the vaginal wall causing a large recto-vaginal tear.

4. The foal has been delivered normally up to the point where the front legs, head and thorax are out, but, in spite of considerable effort by the mare, no more progress is made. In these cases, the hips have either become jammed in the mare's pelvis or, more rarely, the hind legs have flexed and the hind feet have become wedged into the pelvis together with the foal's trunk.

Diagnosis and correction of the second condition is very difficult and should be left to the expert but the first condition can be solved with a little helping traction. This should be applied when the mare forces and should be exerted in the direction of her hocks. The foal should be twisted at the same time. (A major rotation of the front of a foal only rotates the hips a little, but this may be enough to release the hip lock.)

Other, more serious, malpresentations should be attended to by your veterinary surgeon. However, it helps to be able to recognise the serious ones so that no time is wasted in futile attempts at correction.

If nothing is visible, and nothing can be felt in the vagina after ten to twenty minutes of the start of the second stage, then expect problems. The forelegs and/or the head might be deviated laterally, or the foal's trunk might have engaged the pelvic inlet (the breech presentation). Occasionally, the legs and neck can be recognised but are in the wrong position and feel rigid with no normal joint movement. This condition, described as hyperflexion and ankylosis of the lower limb joints is serious and normally a caesarian section is required to resolve the situation.

Posterior presentation, when the hind legs come first, is rare in the horse. It can be recognised by the presence of a tail and the shape of the hocks in the mare's pelvic inlet. Be careful here, as, in the heat of the moment, the hocks can be easily mistaken for the elbows. The feet are usually upside-down in a posterior presentation. This is one situation where speedy, vigorous traction is required as the cord quickly becomes compressed between the foal's chest and the mare's pelvic girdle and the resultant lack of blood will compromise his health.

In conclusion, don't worry about your foaling mare. Consider it a privilege if you are lucky enough to be present during the birth. Keep quiet and try not to interfere unless things start to go wrong. If they do, examine her and make up your mind what is happening. If you are unhappy, then get in touch with your vet. When you call, it will help if you can describe what you have found.

9 Care of the Mare and Foal

If you are breeding your first foal, the day he is born will be a day to remember. Even if you miss the actual birth – and if all goes well, you are very likely to – the sight of a new-born foal finding his balance on extraordinarily long and wobbly legs, then shakily following his instinct in search of his first meal, never seems less than miraculous. Then the speed with which he gains co-ordination and strength is astonishing. His ability to run is the result of his heritage as a creature of flight and wide open spaces. At birth, the foal's reactions are purely instinctive. From then on, however, the learning process begins.

As a general rule, there should be the least possible interference between the new-born foal and his mother during the first hours of his life. Once the foal is safely delivered, it is advisable to dress the stump of the umbilicus with an anti-biotic spray and many studs give a routine antibiotic injection to ward against infections. Otherwise, keep a close eye on the mare and foal, but stay out of their box, or at a distance in the field, to give them a chance to sort themselves out naturally.

Suckling

The foal should be making efforts to stand within minutes of birth and should be on his feet and attempting to suckle within an hour. A mare who has never previously had a foal will often prevent the foal from suckling at first. She may be so over-protective that she will not allow the foal out of her sight and thus keep turning her quarters – and the food source – away from him. Alternatively, she may simply be ticklish and sensitive and refuse to give the foal a chance to find the teat and stimulate her udder to let down the milk.

In either of these cases, help is needed. If the foal was born outside, it will be more practical to bring him and his mother into a loose box, where the mare can be more easily restrained whilst the foal suckles. Hold the mare in a head collar along one side of the box, with a second person to guide the foal in the right direction. The foal will have an instinct to seek a dark space between two uprights and may often mistake any similar object for the milk supply. However, with a little patience on the part of the owner, most foals will find the right place for themselves if mother is held stationary.

Should it be necessary to guide the foal, the handler should avoid putting himself between the mare and foal, but should place one arm around the foal's quarters, to gently push him towards the teat, whilst supporting his chest with the other arm, to prevent him from falling over. Care must be taken to avoid the mare's hind feet (which should be unshod), if she is ticklish and inclined to kick out.

The sucking reflex is instinctive to the new-born foal, but may be lost if several hours go by without him getting his first

Fig 75   Rearing an orphan foal is a job for the dedicated.

meal. The mare's first milk – the colostrum – contains vital antibodies which enable the young foal to fight against infection. If, for some reason, the foal is too weak to stand and suckle, the mare should be milked and the foal bottle-fed. Your veterinary surgeon will show you how to do this if you are in doubt. In any situation where the foal does not show normal, healthy, lively reactions, or is not on his feet and suckling within an hour or two of birth, seek your vet's advice without delay. A sick or weak new-born foal needs immediate professional attention, otherwise irreparable damage will be incurred and death may quickly follow.

The horse's digestive system is designed for continuous action, and the young foal will take only small amounts of milk at first, but frequently. Thus, if the mare is proving difficult about allowing the foal to suckle, help should be provided every two hours, until the mare stands quietly for her suckling youngster and the foal is strong and active enough to persist in getting to the teat. Similarly, bottle-reared foals must be fed on a two hourly basis, day and night, for the first few weeks. Thereafter, the night-time feeds can be gradually reduced and the period between the day-time ones lengthened to three hours. Hand rearing a foal is a job only for the dedicated.

109

Proprietary dried mare's milk substitutes are available and are the best alternative to the mare's own milk if you have to bottle feed a foal. Follow the manufacturer's directions, or your veterinary surgeon's advice on how much to feed.

Should you find yourself in the unfortunate position of having an orphan foal, or a mare who has lost her foal, the National Foaling Bank (*see* Appendix 2) might be able to help, by organising a fostering arrangement with another owner. They can also advise on hand rearing.

Fostering usually involves skinning the dead foal and covering the foal to be fostered with the skin, for twelve to twenty-four hours, having previously starved the fostered foal for an hour or two, which will encourage him to try to suckle from his substitute mother. The mare should be allowed to smell the foal,

with her hindquarters turned out of the way, before the foal is encouraged to suck. The exercise must be repeated often throughout the first day until, hopefully, the mare seems to accept the foal as her own. Thereafter, a close watch must be kept to see that the foal is not being rejected. The success rate for fostering varies and depends very much upon the skill of the handlers, so expert advice should be sought if possible.

Stabling

If your foal was born outside, with no problems and the weather is fine, he and his mother are best left out, in their natural surroundings, for as much of the time as possible. If he was born in the stable, turn the mare and foal out as soon as possible, once you are sure he is

Fig 76 Mare and foal should be turned out as soon as possible.

Fig 77   The foal will spend much time lying down.

suckling properly and all is well. Outside, the mare can relax, roll, graze, walk around and recover from the birth in a natural way. The foal will have space to learn to use and stretch his long legs and will soon start to play. He will also spend a great deal of time lying down and sleeping, especially after a meal, with his mother grazing nearby.

Check on the foal every three or four hours for the first two days. Foals thrive in mild or warm, sunny weather, when the grass is growing and there is just sufficient breeze to keep the flies away. Unfortunately, such weather cannot be ordered to coincide with the birth of your foal and you must maintain a flexible routine to cope with the vagaries of the climate. In a good spell, there is no need to bring the mare and foal in at night and a light shower will do no harm. Provided mother is around to provide a constant supply of warm milk, the healthy foal can also stay out on chillier days without ill effect. However, most studs do not like foals to get their backs wet and will bring them in rather than take any risk of them getting chills, or worse, pneumonia. It is as well to follow this example if heavier or persistent rain is forecast.

One good reason for bringing your mare and foal in for an hour or so each day is to teach the youngster to accept human contact, to become accustomed to

Fig 78   Early human contact is essential. This foal, out of a Thoroughbred mare, is by Tarim (see page 26)

being handled, and to lead. What he learns in the first few days of life will not be forgotten and the right start will save you much time and trouble later on. The best time to begin is the day after the foal is born, when it will be easy to fit a foal slip (a specially designed small head collar) with a second person to calmly restrain the foal in the usual way. Be sure that the foal slip fits correctly – a foal will often reach a hind foot up to scratch, in annoyance at the foal slip, or simply to scratch behind an ear, and his tiny foot can easily slip between his head and a foal slip which is not snugly fitted. There should also be ample room for adjust-

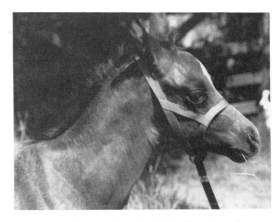

Fig 79   A foal slip should be fitted in the first few days of life.

112

ment, which will soon be necessary, as foals grow surprisingly rapidly.

Some studs, who have purpose-built fencing and many foals to cope with, prefer to fit small head collars which are left on all the time. For the private owner, with one or two mares, this is best avoided, but if you must do it, use leather head collars as opposed to webbing ones and be *absolutely sure* that your fences are safe, with nowhere that a foal might become caught up – a broken neck could easily be the result.

Leading

Taking your mare to and from her stable is a useful time to teach the foal to lead. He will, of course, follow her without being led, but this can also give him the opportunity to get into mischief and, in the long run, it will be safer for him and less trouble for you, if he learns to lead happily from the start.

Two people are needed – one to lead the mare and the other to cope with the foal. To begin with, it might be helpful to

Fig 80   Teaching a foal to lead. The handler has one hand around the foal's hindquarters to encourage him forward, although this foal is not at all reluctant.

113

have a third person, to walk behind the foal, open the gates and generally assist, especially if the foal is big and boisterous or the mare is of a nervous and excitable disposition. This can occur with some mares, particularly Thoroughbreds, if they are used solely for breeding and are not frequently handled.

Choose a safe place to teach your foal to lead – you can begin to give him the idea in the loose box, provided there is enough space, in a straw-covered barn, school arena, or small paddock. You will need a long lead line, which is slipped through the ring of the head collar, not clipped to it, then held double in your left hand. If the foal should escape, the lead line will then fall free, instead of becoming entangled in his legs to cause panic.

Place your right hand around the foal's quarters and gently push him forward. The foal will automatically jerk back as soon as he feels tension on the lead line. Be prepared for this. Slacken the line immediately, then gently ask him to move forward again. Use your voice to encourage him right from the start and maintain the hand pressure from behind. After a few attempts, provided you remain calm and patient, the foal will begin to accept the idea. Do not continue for more than a few minutes, the first time, as a young foal is very easily tired.

Once you are outside the stable and have more space, you can begin to teach the foal to move forward beside his mother without continuous pressure on his quarters, but be prepared to provide it instantly, to encourage the foal, if necessary. Some people like to have a soft webbing line, passed around the foal's quarters and held over his back in the right hand, which can be tightened instantly, when needed. Pushing the foal forward from the rear avoids both the need to pull on the head collar and the likelihood of him jerking right back and possibly going over backwards in an effort to get away from the restricting lead line.

If you do find yourself in front of the foal with him tugging back violently, slacken the lead line immediately – he will stop pulling as soon as there is nothing to pull against. Re-position yourself (a second person to help is advisable with a bigger foal), and take up the line again, being prepared for the foal to pull back immediately he feels the pressure. Urge him forward using pressure from behind. Never wave a whip or other object, nor shout at the foal to make him move. You will only succeed in frightening him. Patience and calm are essential to give him time to understand and learn what you want. If your foal does rear up and go over backwards, little harm is likely to be done, provided he is in a safe place, with a soft landing and you release the lead line immediately. However, there *is* a risk of injury if the foal flings himself around violently. Tactful handling can do much to avoid this.

The more your foal is handled, the easier it will be to train him later. This should include scratching his poll and withers – something that horses often do for each other – reaching an arm around his body and under his girth, where later the saddle and girth will go, running your hands down his legs and picking up his feet. Repeated regularly, these small actions will soon teach him to accept human contact and that he has nothing to fear from you and the strange things you do to him.

The Mare's Diet

Once the foal is born and your mare is feeding him, what should you do about her diet? The simple answer is, if she looks well, is carrying enough condition and has a sheen on her coat, change nothing. Many people worry about increasing the mare's protein intake to supply the foal's needs, but if you have been feeding stud cubes, or a cereal-based ration with high-protein balancing nuts during the last third of gestation, this, combined with the growing spring grass will provide all the protein that is required. In any case, private horse owners do not usually have readily available facilities for testing the protein content of various feedstuffs, and trying to work out a strict diet on a percentage basis is expensive and impractical. Traditionally, experienced stud men have fed mares and foals by reference to nothing more scientific than an observant eye. Anyone with enough horse knowledge to embark on the uncertain business of breeding a foal should be able to assess a horse's nutritional requirements, and, indeed, the horse's general condition and state of health. If you do not possess this much knowledge and confidence, leave breeding to those with more expertise.

Hay, of course, can be discontinued once grass growth is under way and most horses will start to ignore their hay ration once the grass provides enough nourishment. Hard feed should be continued whilst the mare is lactating, except for hardy breeds such as native ponies who tend to run to fat on anything but rough pasture as soon as the new grass comes through. The quantity of feed required depends, as always, on many factors – breed, size, age, condition, the amount of milk the mare is giving and whether she is inclined to 'do herself better than her foal'. You may need to increase the quantity of hard feed for a mare who gives a great deal of milk in order to prevent too much loss of condition. Some mares, however, will not increase their milk supply no matter how much food they are given, using the additional nourishment to grow fat themselves. Care must be taken not to overfeed these mares and if your mare starts to put on weight, decrease her ration. Foals born to mares in this category benefit particularly from supplementary feeding.

The Young Foal's Diet

Young foals have a powerful sense of curiosity, which may lead them into trouble, but which is also vital in teaching them how to survive. From the first few days, the foal will attempt to graze, by copying his mother, and during this process, he will sample the taste of many potential food sources. He does not automatically know which plants are good to eat, but must learn by trial and error from the texture and the taste. His experiments will undoubtedly include sampling droppings and soil and later, chewing bark and twigs and probably the fence, gate or stable door.

This curiosity can also be used to encourage the foal to share his mother's meal times. If you are feeding the mare separately from other horses, it is easy enough to start by offering the foal a little of the mare's ration from your hand. He may simply nose around the bucket or feed bowl when the mare has finished with it, picking up any missed grains of food. As soon as he shows interest in

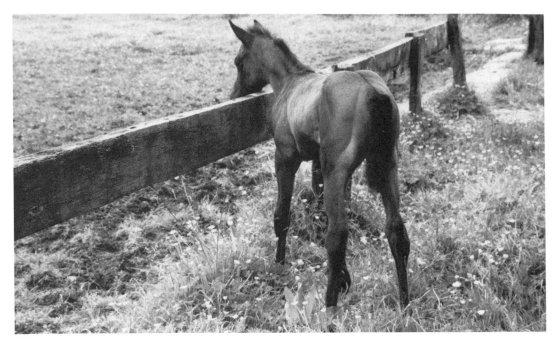

Fig 81   *A powerful sense of curiosity.*

Fig 82   *In early attempts to graze, the young foal must bend his knees to reach the ground.*

116

Fig 83   The foal will watch and attempt to copy his mother.

Fig 84   Experimenting with tastes and textures – including the fence.

nibbling at the food, provide a separate container for him and give him a small ration of the same feed. The foal's feed bowl should be off the ground to begin with, as many young foals have some difficulty in reaching down to ground level. Make sure that the bowl is secured to some safe, immovable support. At first, the foal will eat only small amounts and slowly, but by the time his mother has finished her feed, he will probably have eaten enough to satisfy his appetite. (Proprietary nuts may be too hard for the youngster to cope with at first, but they can be softened by mixing the whole feed with a little soaked sugar-beet pulp and dampening it.) Any left-over food must be thrown away, not offered again at the next meal time.

Since foals cannot manage more than small quantities of food at once, some studs like to provide creep feeders, so that a continuous supply is available. However, if you have just one mare and foal and do not want to bother with this system, feeding the foal whenever the mare is fed will be adequate.

The Mare's Future Use

At some point after the birth of the foal, if not before, the mare's future must be decided. Is she going back to stud to breed another foal next year? Is she going back into work? Or is she going to be left in peace to bring up her foal at home until weaning time? Do you want to show her with the foal?

If you want to continue breeding from

Fig 85   Foals lined up for judging. Wait until your foal is old enough to cope with the stress of a day away from home.

the mare and it is necessary for her to travel away to stud, it is preferable to send her before foaling, so that the foal will be born at the stud and will not have to cope with what may be a long journey within a few weeks of birth. Similarly, if you intend to show your mare and foal, wait until the foal is strong enough to cope with a long day away from home. Most bigger shows require a foal to be at least one month old before appearing in the show ring and, in any case, a bigger, well-grown foal has more chance of success than a smaller, younger one. Also, remember that exposure to strange horses increases the risk of the foal picking up infectious diseases.

Foals tire quickly and the combination of travelling with the excitement, bustle and noise of a show can be very stressful, so if you do show your mare and foal, be well prepared and organised and avoid showing too often. Before travelling your foal anywhere, accustom him to loading into your horsebox or trailer. The earlier you do this, the less trouble there will be later on. Provided the mare is good to load, there should be no difficulty in loading the foal.

It is possible to bring your mare back into work, if you wish, a few weeks after the birth of the foal, though many owners prefer to wait until weaning time. Mares and foals become upset at being separated and for the foal's own safety he will need to be shut up in a stable whilst his mother is away – a traumatic experience, although the company of another foal will help. In any case, he will need to suckle at least every three hours, so mother cannot be away for long. However anxious you may be about starting riding again, trying to work an anxious mare may not be worth the effort. There is much to be said for leaving mare and foal peacefully together in their field, until the foal is six months old and ready for the next stage of his life.

10 Health of the Lactating Mare and Foal

Normal Development of the Foal

One of the most difficult things to do after the birth of the foal is to prevent yourself from rushing in and checking that the new arrival is normal. Every movement seems to be a sign that something is going wrong which needs our immediate attention and it is difficult to judge when to interfere and when not. A knowledge of the normal development of the foal in those first few hours will help us in our quandary.

Breathing should start within thirty seconds of birth. In fact, the foal often starts to breath once the head and thorax have been delivered. To begin with, the breathing is gasping in nature and is accompanied by almost convulsive movements of the head and forelimbs, but, within a few seconds, it has settled into a fast but regular respiration rate of 70/min. The convulsive movements of the head and forelegs should have broken the amnion. If they have not, then we should interfere for the first time and break the membrane so that the head is freed and the foal can start to breathe.

Over the next ten minutes the foal's reflexes develop. The seemingly random movements which a new-born foal makes become more purposeful and he soon rights himself and assumes a sitting position, with hind legs drawn up under his body and forelimbs outstretched, sup-porting the shaky head and chest. The eyelids start to blink and the lips make sucking movements. He may attempt to suckle any convenient object, even at this early stage.

During the next thirty minutes many attempts will be made to get on all four feet and although most end up in an undignified heap on the floor, eventually the task is managed and the first hesitant steps are taken. The foal should be on its feet within one hour of birth and by then should have found the udder and its first meal.

What can go wrong at this stage? I would be a little worried if the foal was not on its feet and sucking within two hours as the all-important antibodies which the mother's colostrum contains can only be absorbed in those first vital hours after birth. The immunity which they provide protects the foal through the first few months of life and without adequate levels, neonatal disease will become a serious threat. An inability to rise might also indicate some other problem – perhaps a locomotive abnormality or a case of neonatal maladjustment syndrome.

Once the problems of the first suckling are over and the foal has had his first meal, the ease with which the foal suckles should improve and the frequency of suckling should increase. The normal frequency should be about once every thirty to forty-five minutes.

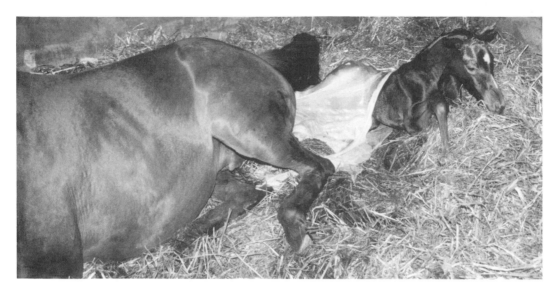

Fig 86   The foal has now righted himself and is able to sit up.

Fig 87   The first attempts at getting up.

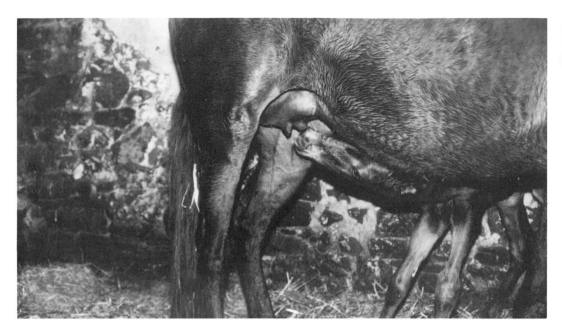

Fig 88   Suckling should soon become an easier matter than this.

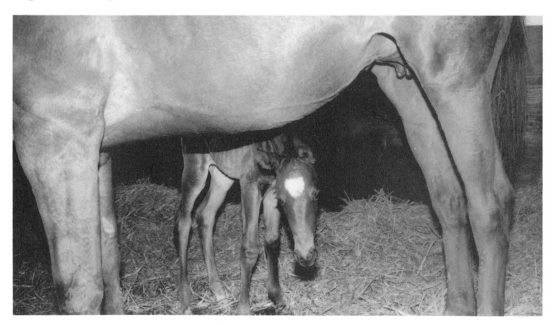

Fig 89   The foal will become an individual, learning new experiences every day but this one is not sure of the big wild world out there.

Over the next few days, the foal should get stronger every day, changing from a very wobbly, rather thin individual, into a confident mover able to keep up with his mother. The angular new-born should have changed into a co-ordinated, more rounded individual who has started to leave his mother's side to explore the immediate surroundings. He should be full of curiosity and we should use this character trait to start developing the bond which we want to grow between him and us. Lots of human contact now will make it easier when formal training starts.

Common Diseases in the Foal

It is inevitable that any chapter on the many and various diseases which a foal might catch is going to read like a mail order catalogue – a long list of items, each with their own short description. To try to avoid this, I am going to group the diseases under three main headings according to the age of the foal. Unfortunately disease is no great respecter of age and many conditions can affect more than one age group, so my grouping will place each disease where it most usually causes trouble. Each group will then be subdivided into infectious diseases and non-infectious conditions.

CONDITIONS AFFECTING THE YOUNG FOAL, FROM BIRTH TO SIX WEEKS

Infectious Diseases

Sleepy foal disease (shigellosis) This is an acute, rapidly fatal disease of the young foal caused by an organism called actinobacillus equuli. The organism is found in the kidneys and brain of the infected foal. The immediate signs include loss of appetite, fever and a rapidly increasing depression which culminates in a coma. In the early stages, the foal can be roused from this coma and even tempted to suckle, but very quickly sinks back into a coma. Hence the name 'sleepy foal disease'. The condition is fatal and even if aggressive treatment succeeds in preventing immediate death, a severe infection of the kidneys and joints will develop over the next few days which will kill or cripple the foal.

Other septicaemic conditions Other organisms can cause a septicaemia that is just as severe. The bacteria that are most often incriminated are streptococcus zoo-epidemicus, some strains of staphylococci, E.coli and Klebsiella. These organisms are common pathogens of the horse and we might ask why they suddenly choose to infect the new-born foal, which they do through the navel cord and by mouth. The answer is, as always when dealing with any disease of the newly born, an inadequate antibody protection against the bacteria invaders. The foal is born with hardly any protection, which then has to be boosted by the first feed of mother's milk with its high levels of antibody. Until the antibody has been absorbed into the foal's system he is vulnerable to all the bacteria which invade his body. This is why it is so important that the foal sucks in the first hour of life.

The first signs of septicaemia are very similar to those of sleepy foal disease and generally occur during the second day of life. The foal becomes slightly dull and disinclined to get up, he does not feed with his usual enthusiasm and frequently allows the milk to dribble all over his muzzle. His temperature will start to climb, reaching 41°C, but may fall as the condition worsens.

Treatment of all types of septicaemia must be immediate. Any delay and the organisms responsible can cause lesions in the joints and other organs which are much more difficult to cure. Suitable antibiotic preparations should be given at regular intervals through the day. Good nursing is essential. The practice of administering routine antibiotics for the first three days after birth will help to prevent septicaemia, especially in those foals that may have low antibody levels or are born on premises which are known to have trouble. Prompt dressing of the umbilical stump with an antiseptic solution is essential and foaling boxes must be thoroughly cleansed and disinfected after each foaling.

Septic arthritis (joint-ill) This disease occurs at any time between birth and three months of age and although it is a common sequel to septicaemia it can also exist as a separate entity. The causal organism almost always enters the body through the umbilical cord, where it causes an abscess. From this reservoir of infection the bacteria spread to and localise in one or more joint or tendon sheath.

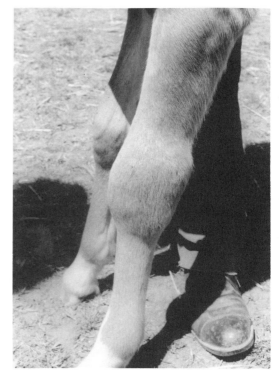

Fig 90   A case of joint-ill.

The first sign of joint-ill is lameness, generally accompanied by swelling of the affected joint. The joint becomes hot and painful, body temperature rises to between 38.5 and 41°C and the foal stops sucking.

As with all these conditions, the state of the mare's udder often gives a better indication of the foal's health than the foal itself. Always check a foal carefully if the mother's udder is swollen, or if milk is escaping from the teat – it generally means that the foal is ill.

It is essential to start treatment of joint-ill immediately. Any delay will mean that the damage done to the joint surfaces and the bone will be irreparable. A long course of an effective antibiotic given early on will sometimes cure the condition. Local treatment of the joint is essential. The joint should be flushed with saline and antibiotics introduced into the joint capsule. Unfortunately, even with immediate treatment, the prognosis must always be guarded as many cases end up chronically lame.

Diarrhoea Enteritis is a common condition in foals of all ages and can be caused by many agents. However, one period in the foal's life when diarrhoea often appears is during the foal heat. The mare comes into heat eight or so days after foaling and during or just after this heat some change occurs which can cause diarrhoea in the foal. The diarrhoea is transient in nature and as long as the foal looks well and keeps on feeding then little harm is done and no treatment is needed. In these cases, a little common sense is required. Does the foal look ill? If he does then treatment will be necessary and you should call your vet. The cause of this diarrhoea is, as yet, unknown, but it may be due to the increase in oestrogen level in the milk.

The round worm Strongyloides westeri can cause a persistent yellowish diarrhoea in the three- to six-week-old foal. For a long time, the ways in which the Strongyloides larvae infected the young foal were not understood but the discovery of larvae in the mare's udder and milk led to the understanding that the method of infection was from larvae produced by adult worms which lived in the udder of the mare. These can infect the foal via the milk as soon as four days from birth and develop full potency during the next two weeks. The eggs can be seen in the faeces from two weeks post birth, reaching a peak at two to three

months of age and disappearing at six to nine months when a self cure occurs. In favourable conditions, larvae can infect the foal through the skin.

Apart from a transient scour this worm causes very little trouble but heavy infections can cause a chronic diarrhoea and loss of weight. Treatment with the anthelmintics Thiabendazole (Thiabenzole) or Ivermectin (Eqvalan) is very successful.

In the slightly older foal, a chronic diarrhoea may be caused by too much milk and it is always worthwhile restricting the milk supply or even weaning early in such cases. If the foal looks ill or, as sometimes happens, the diarrhoea becomes persistent, an enteritis caused by E. coli or Salmonella species may be the reason. A severe type of enteritis can also be caused by a rotovirus or coronovirus infection. Most cases will respond to treatment with replacement fluids and oral antibiotic.

It is important to realise that, rather than the damage done to the intestinal mucosa, it is the loss of fluid from the foal's body that does the harm. The dehydration caused by this loss and the fever caused by the bacteria stop the foal sucking, making the dehydration even worse. In these cases, it is as important to correct the dehydration as it is to treat the enteritis. This is done by administering electrolyte fluids, made up of a mixture of salts and water (approximating to the composition of the foal's own body fluids), by mouth, or, in more serious cases, by intravenous injection. Once the causal organism has been identified, the necessary antibiotic should be administered.

Equine herpes virus type 1 This unpleasant virus causes most problems in the older horse, either as a respiratory condition in yearlings or as an important cause of abortions in mares. It is in this latter context that I mention it now. An infection with EHV1, late in pregnancy, will often severely affect the foetus but not kill it. In these cases, the foetus is not aborted and the foal is born to term alive but very weak and ill. The virus travels across the placental barrier and invades the foetus, severely damaging the liver and other organs. The new-born foal does not survive for long.

Non-Infectious Diseases

There is a whole group of conditions which are non-infectious in nature which can affect the future of the foal. Most of these develop during the inter-uterine period of the foal's life and some are quickly fatal whilst others either cure themselves or respond to some type of treatment.

Developmental abnormalities   The causes of those conditions where various parts of the foal's body have not developed correctly are not fully understood. Certainly, some conditions are due to hereditary factors and we know that various drugs have a tetragenic effect upon the foetus and that inter-uterine infection can cause developmental defects. As a general rule, it is as well to avoid long treatments with any chemical agent during the first six weeks of pregnancy.

Hyperflexion of limb bones   Hyper-flexion or contracted tendons is the most common abnormality found in the foal. The cause is unknown but the position which the foetus adopts in the uterus and the space available has an influence on the development of the condition. The fet-lock is the joint most commonly affected but the knee can also show signs.

Most cases cure themselves as the foal gains in strength and starts to exercise himself, but physiotherapy applied to the joint will help. Five or ten minutes every day spent straightening the joint will speed recovery. More serious cases will have to be splinted in an extended position. The splint needs to be in place for two weeks and great care must be taken to prevent any pressure sores developing. Good padding is essential. It is debatable whether extreme cases (those where the flexion cannot be reduced manually) should be treated, as it is unlikely that a return to full athletic ability will ever be reached.

Hyperextension of limb bones   In this case, the reverse occurs and the foal is born with fetlock joints in full extension. Usually, all that needs to be done is to protect the pastern joint from damage and the condition will right itself as the foal gets stronger. Refractory cases respond to a heel extension which is best applied using one of the new plastic glue-on shoes.

Ankylosis of the joints   This is a condition which carries a much more serious prognosis. One or more joints are fused in a flexed position and cannot be manipulated in any way. Because of the resultant shape of the foetus, the condition often causes severe problems during birth and is a common cause of dystocia (difficult birth). If the foal is born alive, then, even if only one joint is affected, euthanasia should be considered as there is no practical treatment.

Abnormal facial bones This group of conditions affecting the facial bones of the foal is generally considered to be hereditary in character.

1. Parrot jaw. As its name suggests, this condition is characterised by an upper jaw (maxilla) which overhangs the lower (mandible) by as much as three to four centimetres. The incisor teeth do not meet and the molars overshoot at the back and front. Slight abnormalities tend to correct themselves.
2. Squiffy face. The reverse happens in this case. The bones of the upper jaw are much shorter than the lower jaw. Considerable twisting also occurs in both bones. The result is a squashed-up face with a twisted muzzle. The tongue often protrudes and in severe cases the foal finds it almost impossible to suckle.
3. Cleft palate. The first sign of this condition is a constant dribble of milk down the nose whilst the foal is suckling. A gap in the bones of the roof of the mouth allows milk to regurgitate down the nostrils. Surgery will be necessary to repair the defect.
4. Hydrocephalus. The enlarged head and excessively dome-shaped forehead that signify this condition often cause dystocia. Luckily, most cases are born dead, thus avoiding the unpleasant task of euthanasia.

Soft tissue abnormalities The following group of conditions all affect the soft tissues and although some are caused by the trauma of birth and others are developmental faults they all take a few days before they are noticed.

1. Atresia coli. Some part or all of the large intestine fails to develop properly and although the foal appears normal at birth, signs of low-grade colic which gets rapidly severe, are soon apparent. Surgery is needed to confirm the diagnosis and sometimes the faults can be repaired.
2. Ruptured bladder. The cause of this condition is probably excessive trauma during birth. A defect occurs in the wall of the bladder and urine escapes into the abdominal cavity. The symptoms mimic that of a much more common condition – that of retained meconium. The foal strains in an attempt to pass the urine trapped in the abdominal cavity and tends to roll and lie in abnormal positions. The abdomen becomes distended with fluid and except where the tear is very small, urine is passed in small quantities.

The condition is diagnosed by performing a paracentesis (a procedure involving the tapping of fluid in the abdomen), to demonstrate the presence of urine which can be recognised by its smell and chemical constitution. Surgery to repair the defect should be performed as soon as the condition is recognised and the outcome is generally good.
3. Urachal Fistula. During the interuterine period of the foal's life, the urine, which is produced by the kidneys, is drained through a tube called the urachus into the allantoic space. Sometimes, this tube does not close as it should in the first few hours after birth and urine is allowed to escape through the urachus and umbilical cord. The area becomes necrotic and infected and thorough cleaning and cautery of the umbilicus is necessary to encourage healing. Those which do not heal will require surgery.
4. Hernia. There are two common types which, although present in the older foal, are developmental abnormalities and can be considered here. A hernia

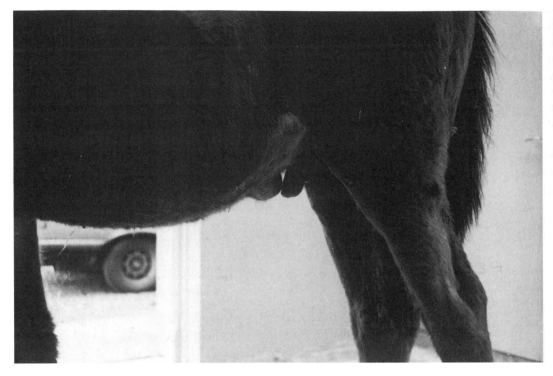

Fig 91   An umbilical hernia in a yearling.

is caused by a defect in the abdominal wall, either in the umbilical or inguinal regions. The peritoneum (the lining of the abdominal cavity) invaginates into the gap and a varying amount of abdominal contents follow. In the case of the umbilical hernia, this is seen as a round swelling, varying in size, which appears at the umbilicus during the first six weeks of life. The hernia can be differentiated from an abscess because the soft abdominal contents can be returned to the abdominal cavity with the application of gentle pressure and the edges of the ring can then be felt.

Unless the abdominal contents become constricted at the ring and strangulate then umbilical hernias can be left as most will regress during the first year of life. If the contents should strangulate or if the hernia must be repaired for cosmetic reasons then surgery is necessary. When the reason for repair is cosmetic there is no hurry, but obviously a strangulated hernia must be dealt with immediately.

An inguinal hernia is a potentially more serious condition. In this case, the abdominal contents herniate through a weakened or enlarged inguinal ring and are seen as a swelling in the inguinal region. The hernia is often involved with the scrotum and, in these cases, appears as a grossly enlarged scrotal sac. Inguinal hernias can be unilateral or bilateral and do not regress, indeed inguinal hernias tend to get larger with age and are more likely to strangulate. Surgery should be performed as soon as is practicable.

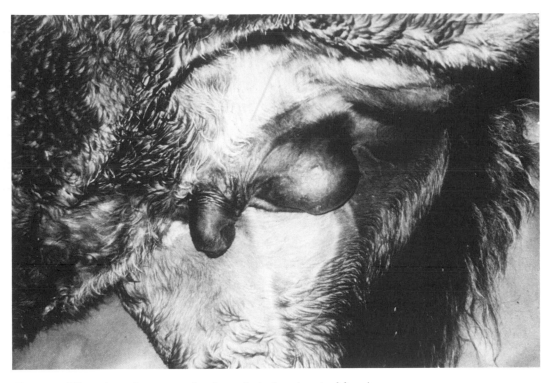

Fig 92   The enlarged scrotum that is typical of an inguinal hernia.

Immunological conditions   Haemolytic disease is the name given to a rather rare condition of the new-born foal involving the blood system. The red blood cells of the infant are destroyed by specific antibodies which the foal absorbs from its mother's colostrum during the first thirty-six hours of life. The antibodies are formed by a mare whose body assumed that the small amount of her foal's blood which leaked into her body during pregnancy was foreign and to be challenged. The destruction of large numbers of the foal's red blood cells causes a severe anaemia, which, if untreated, will cause death within a few days.

The symptoms of haemolytic disease occur within forty-eight hours of birth.

The first signs are rapid breathing and an accelerated heartbeat due to the anaemia, and yellowing of the whites of the eyes and mucous membranes due to jaundice. In the early stages, the foal yawns frequently and becomes increasingly sleepy. Blood-stained urine is frequently present. Treatment consists of removing the foal from the source of the problem, the colostrum, and adding to or better still replacing the foal's red blood cells by transfusing the foal with a new blood supply.

Once a mare has been sensitised to her foal's red blood cells and has generated these antibodies, she is likely to do so again in subsequent foalings. Her future foals can be protected from this disease if

129

they are prevented from suckling for the first forty-eight hours by muzzling or by removing them from their mother. The necessary colostrum can be given to them from another mare.

Behavioural conditions Three conditions which are non-infective in nature and which belong in no definite classification, have been included in this group because they are all characterised by severe behavioural problems.

1. Retained meconium. Colt foals – especially large ones — can find it difficult to pass the meconium, which is the name given to the faeces which have collected in his rectum during pregnancy. This material, which is blue–black in colour and fairly hard in consistency, is normally passed during the first forty-eight hours of birth and is followed by the normal yellow–orange faeces. The cause of this condition is unknown but vitamin A deficiency during the latter stages of pregnancy might be implicated. The condition is more common in foals who are overdue.

The first signs are those of general discomfort. The tail starts twitching, and the foal spends more and more time lying down, looking around at his flank and straining. The discomfort gradually increases in intensity and the foal is often found lying in an awkward position on his back with forelegs folded over his head. Treatment consists of liquid paraffin, given by mouth or as an enema. This, on its own, will remove most blockages. Care should be taken if manual removal is attempted as the rectum of the foal is a delicate structure and easily damaged. Also, manual removal is rarely successful as in many cases the blockage extends far back, beyond reach, into the colon. As a last resort, the retained meconium can be removed by surgery.

2. Neonatal maladjustment syndrome. Foals suffering from this syndrome start life normally but within twenty-four hours they start to show symptoms which have led to the more common names for this condition. Barker, dummy or wanderer foals, as they are known, might begin by suddenly going into convulsions. Between these fits, which vary from total and continuous to slight twitching of head or limbs, the foal sometimes becomes comatose. Rather than convulsing, the foal might wander aimlessly making strange barking sounds or stand with his head pressed up against a wall or the floor. The suckling reflex is generally absent but he might be able to suck from a bottle, even if in a manic way. The symptoms of this syndrome vary tremendously from the vaguest nervous signs to a complete and total uncontrollable seizure but all affected foals show some subtle behaviour change and all generally lose the ability to suckle the mare.

What causes this condition? It seems probable that it is a combination of events of which the most important is trauma during birth. This can lead to circulatory disturbances and a lack of oxygen which cause or exacerbate existing lesions in the brain. The brain damage which follows results in further circulatory and respiratory disturbances which cause more nervous damage. This vicious cycle which rapidly culminates in death must be arrested.

The treatment concentrates on stabilising the foal so that the original damage can heal and the foal can return to normal function. The nervous signs, if severe,

*Fig 93   This foal, suffering from neonatal maladjustment
syndrome, was found too late and did not survive.*

can be controlled with anti-convulsant drugs or, if necessary, by light anaesthesia. Some foals may only need supervision to prevent them wandering into harmful situations. Food should be administered by stomach tube if the suck reflex is absent and some more severely affected foals might need fluid therapy. As a general rule, all but the mildest cases need the attention of a specialist centre in order to cope with the many complications which this syndrome presents. On the happy side, however, if the condition is treated quickly and the foal stabilised, then most cases recover to lead completely normal lives.

3.   Premature birth. The mare has such a varied gestation length, from 320 to 355 days, that it is difficult to know when a foal is premature or not. Generally those born over 300 days but before 320 days are considered premature. The premature foal can be recognised by virtue of its small body weight (the average normal weight of the Thoroughbred is in the region of 49kg), and by a general weakness and inability to stand. He has difficulty in suckling and in keeping warm and has a thin scrawny look. The hair and skin feel silky to the touch and the mucous membranes and tongue are often brick red in colour.

131

Successfully rearing a premature foal is a very rewarding job but often full of disappointment as he will be extremely vulnerable to all the previously mentioned diseases and, more often than not, succumb to one or other of them. The following are the main points to remember when rearing a premature foal.

a) Make sure that an adequate supply of colostrum is fed within the first few hours of birth.

b) Keep the foal warm at all times. Premature foals seem to be unable to maintain their body temperature.

c) Feed small amounts of food often, throughout the day and night. If the foal is strong enough, this can be achieved by supervising natural suckling but a weaker foal might need bottle-feeding or milk and fluids administered through a nasogastric tube.

b) Preventative antibiotic cover should be provided in an attempt to stop a septicaemia developing.

CONDITIONS AFFECTING THE FOAL FROM SIX WEEKS TO SIX MONTHS

Infectious Diseases

Respiratory infections Young foals are rather like children at their first term of school – always catching colds. Normally, these are of little importance. In fact, where groups of youngsters are kept together spread of infection is encouraged to stimulate a wide immunity.

1. Upper respiratory tract infections. Respiratory infections in the foal are caused primarily by viruses but very

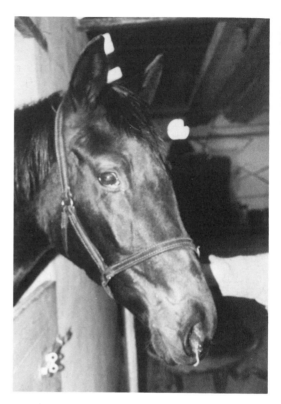

Fig 94 A typical nasal discharge seen during an acute upper respiratory tract infection.

soon a bacterial infection complicates the original infection and causes the typical upper respiratory tract infection which we see in the young foal. The viruses involved are the herpes group, of which equine herpes virus type 1 is the most common, the equine influenza virus, adenovirus and rhinovirus. They all cause much the same initial symptoms which are concentrated in the upper respiratory tract area. A profuse nasal discharge is the first sign. This starts as a watery gush but quickly changes to a thicker, mucopurulent one which clogs and scalds the nostrils. A cough may be present and the foal might look ill for a few days although most affected foals recover within a week to ten days.

Treatment is not usually necessary except to keep the nostrils clean and to apply a healing ointment if scalding occurs. Cases where the nasal discharge persists or where the cough becomes chronic should be treated with antibiotic. One of the mucolytic drugs will also help to clear up the condition.

Foals should receive their first dose of equine flu vaccine at two to three months of age. By the time the course is completed they will be protected against at least one of the respiratory viruses.

2. Lower respiratory tract infection. Infection with one of the influenza viruses or with adenovirus and secondary bacterial infection, generally with streptococcus equi or zooepidemicus, can sometimes lead to a pneumonia which should be treated more seriously. Pneumonia should be suspected when the temperature rises and stays high, when the respiratory rate increases and a harsh cough develops. The foal will become depressed and refuse to suck. Treatment with antibiotic and mucolytic drugs should be instigated immediately.

One particular bacteria called Corynebacterium equi, luckily quite rare in this country, can cause a specially nasty type of pneumonia in two- to three-month-old foals. The bacteria enters the body via the mouth or lungs and is thought to be present in the soil. Migrating worm larvae might transport it through the body. It appears to be concentrated in local areas with particular farms and studs having a high incidence.

The course of the disease is insidious in nature and it is generally well advanced before any obvious signs are seen. The bacteria causes local infection and abscess formation in the lung and, on post mortem, can be identified in large numbers in the creamy pus which is found inside the well-encapsulated abscesses. The disease should be suspected when a harsh cough with concurrent loss of weight and a high temperature that is refractory to treatment, are present. Treatment is unsuccessful. The bacteria is sensitive to many antibiotics, but, by the time the condition is suspected, it has walled itself off in the abscesses which have formed and is beyond the reach of treatment. Very few cases recover.

Before we leave the respiratory diseases, some mention should be made of the cough and running nose which can be caused by migrating worm larvae. The larvae of some worms infecting the foal pass through the lung during their life cycle. As they do so, they cause a transient cough and nasal discharge which generally clears up with no trouble.

Diarrhoea The various causes of enteritis have already been covered in the section on diseases of the foal under six weeks. Suffice it to say that enteritis caused by infection with salmonella and other bacteria can also cause disease in the older foal and any scour that does not clear up within a few days should be investigated.

Tyzzer's disease This is a rare disease affecting foals of about six weeks of age. The causal organism, a baccilus called bacillus piliformis, infects the liver causing an acute hepatitis. The disease is acute, generally causing death after a short illness of one or two days although quite often the foal is found dead for no apparent reason. The symptoms are like those for any acute illness in the foal – severe depression, followed by collapse, convulsions, coma and death. A post mortem is required before the condition can be diagnosed.

Joint infections Septic polyarthritis is not a disease which is found only in the young foal. It can also occur in the foal up to six months of age. It should be suspected when any joint suddenly becomes swollen and painful, especially if this is accompanied by a rise in body temperature and a depressed foal. An infection with one of the salmonella species is the probable cause, and, as with the disease in the younger foal, treatment must be started immediately if joint function is to be maintained.

Non-Infectious Conditions

Combined immunodeficiency disease Combined immunodeficiency disease (CID) is a disease affecting Arab foals. The disease is primarily due to an inability to produce immunity to any disease. While the foal is covered by the immunity passed on from the mare, via her milk, he appears normal, but as the maternally derived immunity declines over the next few months, he becomes susceptible to a variety of diseases of which a pneumonia caused by adenovirus is the most serious. The condition is due to a genetic defect transmitted as a recessive character. There is no cure and most foals die as a result of repeated attacks of respiratory tract infections.

Cerebellar hypoplasia This is another disease, again thought to be caused by a recessive genetic defect, which is found in Arab or part-bred Arab foals. The condition is caused by a degeneration of the cells in the cerebellar region of the brain. Symptoms begin at about four months of age. There is a gradual loss of balance, generally starting with the hind legs but rapidly leading to complete ataxia. The forelegs show an exaggerated high stepping gate and the head nods in a typical, almost shivering motion. The frequency of the attacks increases until the foal becomes permanently recumbent. There is no cure for this condition, and, as soon as a positive diagnosis can be made, humane destruction should be carried out.

Limb deformities In the older foal, most, if not all, limb deformities are the result of the rapid growth and mineral and vitamin imbalances caused by our persistent desire to overfeed young horses. The wish and belief that a fast-growing over-fat foal is 'doing well' is erroneous and the economical pressure to get a young horse to adult weight as soon as possible is no friend to the horse or, indeed, to the owner.

Osteochondrosis Bone grows by increasing its length and diameter at two areas – one called the growth plate and the other, deep in the cartilaginous layers at the ends of the bone. The immature cartilage layer grows in these two areas, thus increasing the length and breadth of the bone. At the same time, the new cartilage changes into young bone. This process occurs smoothly across the whole width of the growth plates to ensure that the bone grows straight and is capable of bearing weight throughout its development. A combination of factors, of which overfeeding, too fast a growth rate, mineral imbalance and over-exercise are the main components, conspire to create an imbalance. The rate of ossification cannot keep up with the rate of new cartilage formation and as a result the deeper layers of the cartilage become necrotic and cannot support the weight of the foal. This condition is called oesteochondrosis and is thought to be the root cause of most deformities of the limb bones of the actively growing foal.

Oesteochondritis dissecans This condition is a development of osteochondrosis where, under the influence of an overfat body and periods of hard exercise, the cartilage layer ruptures and the underlying bone becomes exposed to the joint. The action of synovial fluid on the uncovered bone is very painful. The joints most often involved are the stifle and hock with the shoulder and fetlock joints almost as common. The treatment of this condition is not very satisfactory, the important factors being time, rest and a more sensible diet. Prevention of the condition is theoretically simple but practically difficult as it involves educating the horse owning public to feed their young stock in a more natural way and to forgo that satisfying comment from friends – 'He's doing well, isn't he?'

Angular deformities This condition can be defined as a deviation of the distal portion of fore- or hind limbs from the normal line. The deviation can be medial or lateral. Angular deformities can occur either soon after birth, when they are due to incomplete ossification of the limb bones and slack joint capsules and ligaments, or later in the foal's life when abnormal bone development is the cause.

The first condition is present in most

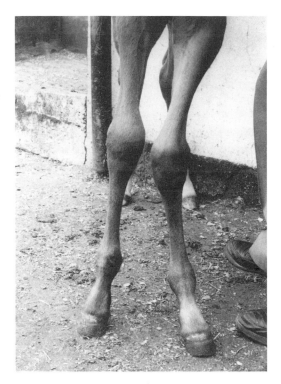

Fig 95   A typical lateral deformity of the carpus.

new-born foals and generally rights itself within the first few days as the foal gets stronger and the joint capsules get tighter. In some foals – the weak, premature ones or those who are forced to exercise too much trying to follow the mare – the condition worsens. These cases must be corrected before continuing ossification makes the angularity permanent. Splinting the leg straight is the best method and is generally very successful. Care must be taken to ensure that the splint is well padded and not likely to cause pressure sores.

As the foal gets older, the long bones increase in length. Most of this increase occurs at the distal end of the bone, in the active growing region called the growth plate. A little growth occurs in the head of the bone. Occasionally, the rate of growth across these regions differs, one side growing faster than the other. Therefore, the leg deviates from the norm. The reasons for this unequal growth are unclear. Certainly, trauma at birth is involved and the unequal forces acting on the joint as a result of developing oesteochondrosis are factors in the development of displacement. The condition is common in the fat, over-done, quickly growing individual which would support the role of oesteochondrosis in this condition – another reason to slow down the development of the foal to a more normal rate.

The treatment of this condition relies upon altering the rate of growth of the deformed bone, either by persuading the shorter side to grow faster or the longer side to slow down. The latter method has been the traditional approach and various techniques have been developed to slow down the rate of growth. These all depend upon bridging the growth plate with a rigid structure – either staples or wire – and preventing any increase in size, thus allowing the shorter side to catch up and straighten the leg. Once this has been achieved, the restraint is removed and the leg allowed to grow on normally. The former method relies upon a recent observation that damage to the covering of the bone (the periosteum) leads to a corresponding spurt in bone growth underneath the damaged area.

The technique involves transecting the periosteum over the growth plate on the shorter side and allowing the accelerated bone growth to level up the leg. This method has the advantage that no artificial material has to be removed once

correction has been achieved and also, due to reasons unknown, once the leg is straight the differential growth ceases.

Whatever the method of correction, it is important to allow enough time for this to occur and to operate during the period of maximum bone growth. In the case of deviations involving the fetlock, this is during the first month of life and for the knee and hock joints, any time during the first four months.

CONDITIONS AFFECTING THE FOAL FROM SIX MONTHS ON

When dealing with the older foal, it is important to realise that many of the diseases already mentioned above (particularly the respiratory conditions), can also infect this age group. Equally, the few conditions that I am going to cover now, although more usually found in the older foal, are also found in the younger age groups.

Tetanus What causes tetanus? The disease is caused by a bacterium called Clostridium tetani which can exist in the soil, as highly resistant spores, for a number of years. When these spores become established in a suitable environment they start to proliferate, and, in doing so, produce a toxin which travels along nerves to the spinal cord and then on to the brain. Once in the brain, the clinical symptoms start to develop.

The organism is particular as to its environment and only grows in a warm, moist situation in which there is little or no air. Deep puncture wounds are an ideal site, particularly in an area in which the blood supply is poor. The type of wound caused by a nail penetrating the

Fig 96   Tetanus – once this stage has been reached, treatment is unlikely to be successful.

sole of the foot or under a burn where a superficial area of dead skin keeps air from reaching the bacteria, are good examples. However, in the vast majority of cases of tetanus, no obvious site of injury can be found. In such cases the horse is probably infected through a small wound in the intestinal wall. This is a very good example of why it is so important to protect your foal routinely rather than wait until your vet suggests it following an accident.

The first signs of tetanus can often be missed as the toxin has a progressive effect. The muscles, particularly of the tail and ears, become stiff causing the tail to lift and the ears to be pricked, but the most consistent sign is a protrusion of the third eyelid across the eyes, which becomes exaggerated if the horse is excited. There is an anxious expression on the horse's face which results from the muscles of the corners of the lips becoming paralysed, drawing them back into a worried smile.

It is at this stage that treatment can still be successful but from now on the disease develops rapidly and prospects become progressively worse. The next stage is for the legs to become paralysed so that the gait becomes stilted and finally the horse develops a trestle-like appearance, unable to move his legs at all. If pushed, he will fall over with his legs still extended. Once the horse is recumbent, treatment is virtually hopeless and the disease develops to cause paralysis of the respiratory muscles and death.

Treatment is directed along several lines but the most important thing is to avoid all stimulation, because the animal is hyper-excitable and excitement causes further muscle spasm and paralysis. This is best achieved by placing the horse alone in a darkened loosebox, keeping noise to a minimum, and visiting him as little as possible. If a wound is present, then the bacteria must be killed by thoroughly cleaning and aerating the wound and applying a suitable anti-bacterial agent. The toxin which has been formed and is acting on the nervous system must then be neutralised. This is done using a tetanus anti-toxin. The best results are obtained when the anti-toxin is introduced into the spinal canal of the horse, getting as near as possible to its site of action. A calming and relaxing agent, generally a powerful tranquilliser, is given every few hours to relax the paralysed muscles sufficiently to allow eating and drinking to take place. The more severe cases might need their food and water to be administered through a nasogastric tube.

Even when a horse is lucky enough to recover from the disease, he remains physically weak for many weeks after resuming normal feeding and movement. It is during this time that expert nursing and feeding is necessary. Tetanus is a terrible disease and despite the best treatment, many horses who develop it die a protracted and very painful death. It seems incredible that so many horses are not protected against the possibility of catching the disease and I would refer you all to the section on preventative vaccination if your mare and foal have not been vaccinated against tetanus.

Skin diseases As the foal gets older, he approaches his first winter and it is during this period that most of the skin conditions that could affect him are seen.

1. Mud fever, cracked heels and rain-scald. These are all conditions which are common in the outwintered foal. All three conditions are caused when the skin becomes infected by an organism called Dermatophilus congolensis. In the case of mud fever and cracked heels the skin of the legs and between the heels becomes red and inflamed, and crusty lesions appear, matting the hair into clumps. The legs swell up, becoming very painful, the tender skin between the heels cracks open, hence 'cracked heels', and the foal becomes lame and sometimes quite ill.

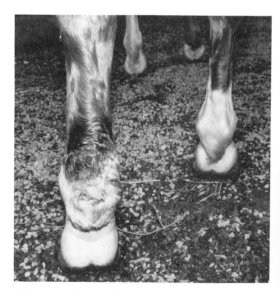

Fig 97   A case of mud fever.

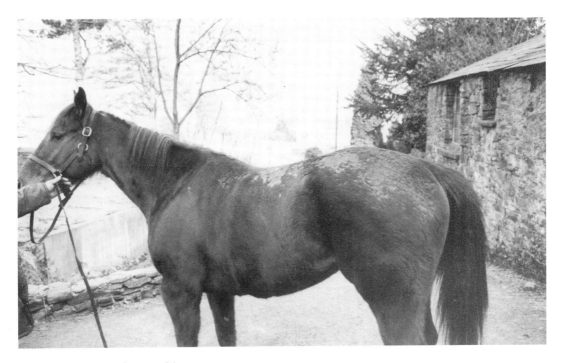

Fig 98   A case of rainscald.

Rainscald is a similar condition affecting the skin of the back, flanks and hindquarters. Again the skin becomes scabby. Under the crusts, the skin is moist, red and inflamed. The extreme irritation makes the foal very uncomfortable and the loss of condition which occurs is marked.

Dermatophilus congolensis is the primary organism associated with these conditions but other bacteria can invade the damaged skin and cause a more severe secondary reaction. Dermatophilus finds it much easier to infect the skin when natural resistance is lowered by the constant wet so characteristic of British winters. The rain washes out the protective skin oils, softens the skin and gives the organism a chance to invade. The mud which plasters the lower legs of outwintered foals can act like sand paper, the small particles rubbing the skin and causing small abrasions through which the organism can pass.

Caught in time, the condition can be cleared up by keeping the skin dry and by gentle grooming to stimulate the body's natural defence mechanism. Any mud on the legs should be allowed to dry and then brushed off. In this way, the abrasive effect of the small soil particles is minimised. If the condition has been neglected then more rigorous treatment will be needed. The skin should be shampooed with a medicated wash and systemic and local treatment with antibiotics will probably have to be used. The hard crusty lesions which develop at the back of the heel and lower leg in cases of cracked heel and mud fever are best treated with an anti-inflammatory anti-biotic ointment. A course of antibiotics is always needed. The infection must be controlled, otherwise the resultant thickening and scar formation will lead to permanent lameness.

2. Ringworm. Another skin condition which is more common during the winter is that highly infectious disease, ringworm. The causal organism belongs to the Dermatophyte group of fungi. The most common is called Trichophyton Equinum but the ringworm associated with cattle – Trichophyton Verrucosum – and dogs – Microsporum Canis – also infect horses.

The lesions start as small areas of raised hair, rapidly becoming crusty and bald. They are generally many in number and are commonly found under some portion of tack. The area under the head collar or at the neck of the rug are favourite places. Ringworm is very infectious and can be carried from horse to horse by contaminated brushes, rugs, numnah and so on. At some time in its development, the fungus forms spores, which are highly resistant and form reservoirs of infection in woodwork, old fence posts and unwashed clothing, which explains why the disease can appear out of the blue.

Luckily, treatment is relatively easy. A seven-day course of an antibiotic called griseovin, given by mouth, will generally kill the fungi, although the hair will take some time to regrow. Very effective skin washes have recently been developed, and, sprayed over the horse every four days for at least three times, these will cure the condition. Be sure to cover the whole body with the wash or some unseen lesion will escape treatment.

Lice   Loss of hair and signs of rubbing should be checked at once. The culprits are probably body lice. They enjoy the comfort of warm long hair and generally start causing trouble from January onwards. To break the life cycle, treatment should be at two-week intervals, either by washing the coat with an insecticidal medication or by dusting with a suitable louse powder. The use of ivermectin worm doser also has the useful side action of killing lice.

Wobblers disease   Wobblers disease is a condition that causes ataxia in the hind legs. The condition is seen predominantly in young male Thoroughbreds but all species can be affected. The first signs can appear suddenly and it seems that the young horse cannot control his hindquarters. At the walk and when turning, the hind legs and sometimes the forelegs collapse underneath the foal and he nearly falls. The slower the pace, the worse the co-ordination.

The condition is caused by a narrowing of the cervical vertebrae and consequent damage to the cervical spinal cord. This is another condition where the underlying cause of the damage is thought to be the condition called osteochondrosis. Wobblers disease is progressive and there is no cure, although we have had a few cases where the initial condition was not severe and the foal could cope with the lack of co-ordination. These cases gradually got better over a period of time.

Health Problems of the Lactating Mare

Mastitis   Mastitis is a rare condition in the mare although it is often confused with the swelling and pain which occurs in the mammary gland when the foal is ill and not sucking. In fact, in cases where the mammary glands are swollen and painful, the first thing to do is to check that the foal is sucking normally. However, mastitis does occur and it is usually caused by the bacteria, Streptococcus zooepidemicus, and it often develops at or around weaning.

The gland or glands become hot, hard and very swollen. The swelling often spreads forward along the ventral body floor and upwards between the hind legs. The whole area becomes very painful and the mare resents any interference, either by humans or the foal. She looks ill and is often off her food.

Treatment consists of parental antibacterial treatment together with local application of an intramammary antibiotic preparation. Local treatment of the mammary gland with hot water applications will ease the pain and might make it easier to introduce the intramammary preparation – often a difficult job as the teat orifice can be very small.

Hypocalcaemia or lactation tetany

This is another rare condition in the mare but when it occurs, it is very alarming and although it responds to prompt treatment, it should be mentioned. Hypocalcaemia, or lack of calcium, is a stress-related illness. The typical times for it to develop are after the mare and foal have been moved by lorry or trailer, when the mare is separated from her foal for a few hours or, very commonly, when the foal is weaned. Only rarely does the condition occur, as it does in other species, just before or after foaling.

The first signs can be confused with tetanus. The mare becomes stiff and lacking in co-ordination, with a glazed look in her eyes. Typically, the mare is found with locked jaws, with some grass or hay protruding from her lips. As the condition worsens, tetanic spasms occur and the mare becomes recumbent.

Treatment is very successful and consists of administering a solution of calcium salts by slow intravenous injection. The mare should be normal a few minutes after treatment.

Routine Preventative Measures

VACCINATION

Vaccination is a procedure where part or the changed whole of an infective agent is injected into the body in order to cause the body to produce antibodies against that agent. An immunity against the disease caused by the bacteria or virus, the infective agent, is then present. The length of time for which the immunity is effective varies between vaccines and all vaccines need boosting to maintain adequate levels of protection. The immunity produced in the mare is capable of being transferred, passively, to the foal via the colostrum, a fact which enables us to start the routine protection of the foal from birth.

The foal should be protected against two diseases as a matter of routine – equine influenza and tetanus – which can be prevented by the use of a combined vaccine or separate vaccines. The usual procedure is to vaccinate the mare with her annual booster a month before foaling, which will pass on enough passive immunity to protect the foal for the first three or four months of his life. Then, at four months, the first dose of combined vaccine can be given, followed by the second dose of the primary vaccination, not less than twenty-one days or more than ninety days after. To satisfy Jockey Club rules, the first of the booster doses of equine influenza should be given not less than 150 days and not more than 215 days after the second of the primary vaccinations. The necessary third tetanus vaccination can be included with the next influenza booster, which is given a year after the first one. Thereafter, influenza boosters must be given every year at an interval of less than a year. Tetanus boosters are needed only every two years.

A vaccination programme against other diseases such as equine herpes and botulism is possible but these programmes are not considered routine and are usually used where there is a high risk of infection.

WORMING

All foals start acquiring internal parasites at a very early age. The first worm to infect the foal is one called Strongyloides westeri which was covered earlier in this chapter and is of little pathological significance excepting in its involvement in foal diarrhoea.

Parascaris equorum A serious parasite affecting young foals is one called Parascaris equorum – the large white worm. This worm is much the same shape as an ordinary earth worm, white in colour, and of a variable size. When they are present in the foal's intestines in small numbers, with little competition, they can grow up to 20cm in length but a more normal size would be 5–10cm. The adult worms are capable of laying large numbers of eggs, up to many thousands per day. This vast outpouring of infective eggs rapidly contaminates the pasture and can lead to very high worm burdens in susceptible foals. Luckily, as the foal gets older, his immunity to infection increases and the number of Parascaris worms in his intestine decreases. The increasing levels of immunity also render the female worms less prolific, so fewer eggs are deposited on the pasture. Parascaris infection is therefore a disease of the young foal under six months of age.

How do young foals get infected? Foals are inquisitive creatures, always licking objects, trying out the taste of grass and nibbling their mothers' hind legs. During this process they inevitably pick up Parascaris eggs which have been passed by other horses. The eggs of Parascaris have three coats – an outer one, very sticky to enable the egg to adhere to hair, grass, boots and stable equipment, and two inner layers, designed to protect the larvae for a long time. Where the outside temperature is below 10°C the egg can remain viable for many years. However, in warmer conditions, during the summer, the egg hatches into an infective larva in ten days, still within its protective coats.

Once the egg reaches the intestine, the larva is hatched out and it passes through the wall of the intestine to the liver. After a few days wandering around the liver, larvae make their way to the lungs where they are coughed up and swallowed again. The larvae continue to develop in the fore part of the small intestine – the duodenum and proximal jejunum. They swim about in the intestinal contents taking in nutrients which the foal badly needs. By eight to ten weeks, they have

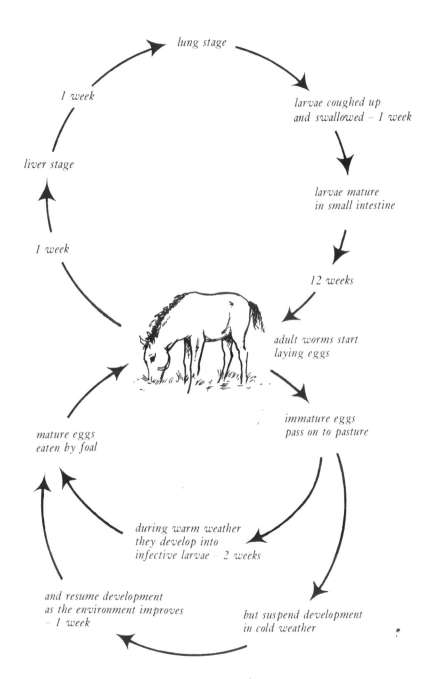

Fig 99   The life cycle of Parascaris equorum.

finished their development, and, now adult, have started to lay eggs themselves.

When large numbers of worms are developing in the small intestine the foal loses his appetite and becomes dull and lethargic. More rarely, the intestines can rupture or become impacted due to the large numbers of larvae present.

How do we control this worm? The first consideration is to try to cut down the numbers of eggs the foal might ingest in the early months of his life. The mare should be wormed two weeks prior to foaling. Her hindquarters and udder should be washed thoroughly after foaling. This will reduce the number of eggs picked up from mother. Foaling boxes should be washed out before foaling to remove eggs from this environment. As far as is practically possible, the pasture on which mother and foal graze should be free of worm eggs. As most eggs are passed out by worms inhabiting foals under six months, this can be partially achieved by ensuring that this year's crop of foals graze on pasture that did not hold foals the previous year. These measures will go a long way towards reducing foal worm burdens but because of the toughness of the egg, its longevity and the large numbers in which they are present, it is impossible to eradicate Parascaris equorum completely.

To prevent the unthriftiness that infection with Parascaris causes in the foal and the occasional sudden death from intestinal rupture or impaction, regular treatment is necessary. This should be done every month or six weeks until the foal reaches nine months of age. By then the naturally developing age immunity will be reducing the numbers of worms present. Although an old standby, the anthelmintic citrazine is still a very effec-tive treatment. The newer ivermectins, given by mouth, are effective against both adult and the younger larval forms of Parascaris equorum.

Strongyles As the foal starts to graze more, it is inevitable that other species of worm eggs are picked up and start to cause problems. The most significant of these are the strongyles. The strongyle family are a large group of worms whose members include the major intestinal parasites affecting the horse. The family contains many types of worms, which can, for convenience, be divided into two groups – the large and small strongyles.

1. Large strongyles. The large strongyles consist of three types. The red-worm (Strongylus vulgaris) is the most common. It also causes the most damage to our horses. The second member of the family, Strongylus edentatus, is not so common and its simple life cycle causes less damage to the horse. The third member of the group, Strongylus equinus, is rarely found in horses in this country.

Red worm has a complicated life cycle which takes seven months to complete. The adult worm lives in the large intestine, browsing on the mucosa (the lining of the intestine). The female lays eggs which pass out on to the pasture and hatch out into larvae, which under warm, wet conditions and going through several changes, become infective larvae in one to two weeks. The larvae are free living and move up the pasture grasses, waiting to be eaten by some grazing horse. Once in the digestive tract, they burrow through the wall of the intestine and migrate along the tiny arteries which supply the gut. They move towards the root of the

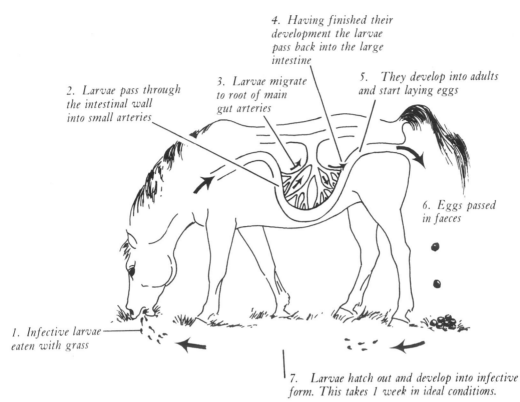

4. Having finished their development the larvae pass back into the large intestine

3. Larvae migrate to root of main gut arteries

5. They develop into adults and start laying eggs

2. Larvae pass through the intestinal wall into small arteries

6. Eggs passed in faeces

1. Infective larvae eaten with grass

7. Larvae hatch out and develop into infective form. This takes 1 week in ideal conditions.

Fig 100 *The life cycle of the large red worm, Strongylus vulgaris.*

main intestinal arteries. It is during this migration that the horse is damaged. This is an important point to remember when understanding how Strongylus vulgaris causes the immense damage that it does. The adult worms cause little harm compared to that caused by the larvae.

During this journey along the fine arteries, the larvae damage the blood vessel and often block the lumen so that blood and, therefore, oxygen is withheld from a section of gut. This is extremely painful and can cause bouts of spasmodic colic. These attacks mostly pass off with appropriate treatment, but they should never be treated lightly, as sometimes the blocked artery is a large one and the

damage to the dependent gut is so severe that surgical treatment is required.

Once the larvae reach the main arteries of the gut (a journey which has taken them about two weeks), they spend three to four months buried in the wall of these arteries. A further migration then takes place as they travel, back along the blood vessels, to the small intestine and from there, to the large intestine. Here they burrow into the wall and spend a further six weeks in hibernation whilst completing their development. Finally, they emerge as adults, when they feed on the lining of the large gut. Considerable areas of mucosa and associated blood vessels are damaged.

145

Strongylus edentatus also has an extended migratory phase during its life cycle, but the path of this migration does not do as much harm as that of Strongylus vulgaris. The larvae pass from the gut, along the portal blood vessels, into the liver. They spend two months developing here, before moving to spend a further few months under the peritoneum (the lining of the abdominal cavity). Finally, they move to the wall of the large intestine and penetrate it, passing into the lumen where they complete their maturation.

Strongylus vulgaris, and, to a lesser extent, Strongylus edentatus damage the young horse in two different ways. On the one hand, we have the damage that the larvae do to the system as they migrate through vital areas of the body, and on the other, we have the damage that the adult worms do as they feed on the tissues and blood of the large intestine.

The migrating larvae cause profound changes in the body, especially when the numbers are high. They block arteries, which, by depriving areas of the body of blood, causes local damage and death of the tissues dependent on that blood. The larvae of Strongylus edentatus can damage the liver as they pass through its substance, and those that lose their way can damage the kidneys and other organs. The damage which results from both types of migration invariably causes rapid weight loss, depression and fever. However, the most serious effect of the larval invasion is the effect it has on the alimentary tract. We now know that a high proportion of cases of spasmodic colic are due to the damage done to the gut following blockage in the blood vessels supplying these areas. In comparison, adult worms are less of a hazard, but if they are present, in large numbers, in young horses, they can cause a chronic anaemia. The damage that the many mouths do to the intestinal mucosa also upsets the fluid balance maintained in the body and causes a diarrhoea.

How do we control this troublesome group of worms? Diagnosing that redworm is causing trouble is the first step. Waiting until we can find worm eggs in a dung sample does not help, as, in the young horse it is the immature worms which do the damage. The changing ratio of certain proteins in the blood is quite a good indication that larvae are causing trouble and a high incidence of spasmodic colic also indicates that larvae are present and causing damage to the blood vessels of the gut. Sudden weight loss is also an indication that worm larvae are present.

2. Small strongyles. The life cycle of the small strongyles is a more simple affair than that of the large strongyles. Adult worms, living in the large intestine of older horses, lay eggs throughout the year. The eggs hatch out in the faecal material and when conditions are wet, they move away into adjacent foliage. Before becoming capable of infecting other horses, the larvae must undergo a period of further development, the duration of which varies, depending upon environmental conditions. In a warm, wet period the larvae mature into an infective stage within a week, but cold, dry conditions slow down the development. This graduated development ensures that the maximum number of infective larvae are present when the conditions are favourable for their survival. Hot, dry conditions kill larvae rapidly, but resistant, newly hatched larvae can survive for long periods inside moist balls of faeces.

Once in the intestinal tract of the horse,

the larvae burrow into the wall of the large intestine and undergo a further stage in their life. This generally lasts for one or two months. The larvae then emerge from their cysts and gradually grow into mature adults capable of laying eggs. Foals can become infected as soon as they start to nibble grass and can have patent infections from two months on.

How do these small strongyles cause illness in horses? As with the large strongyles, the larval stages do most damage. The larvae which encyst in the wall of the large intestine damage the glands of the gut and also affect the motility of the large intestine. Lack of motility of the gut is thought to be one of the more important causes of colic. However, it is during the emergence of the larvae, from the cysts into the lumen of the gut, that most damage is done. As the larvae break out of the cysts and through the tissues of the gut wall, they cause an intense reaction. This phenomenon occurs during the period of maximum migration in the early spring, in the older horse, and if the damage done to the intestinal lining is severe enough, a condition called larval cyathostomiasis results.

This disease is characterised by an acute diarrhoea accompanied by rapid weight loss. The horse generally has a temperature and looks very depressed. Appetite is lost. Without early, effective treatment the prognosis is bad. So, if, in the early spring, your yearling suddenly begins to lose weight and scour profusely, do not wait – call your vet. A new treatment régime, comprising intensive fluid replacement and steroids has seen an increase in the number of horses that survive this serious condition.

Treatment of both types of strongyles depends upon routine dosing with drugs active against these worms, and the period between treatments varies according to the effectiveness of the drug at eradicating the larvae. Thus, monthly treatments, starting when the foal is three to four months of age, are needed when the drugs are active against adults only at normal dose rates. A gap of eight weeks can be left between treatments when a drug which is active against larval and adult forms of the worms is used. Preparations containing ivermectins and those containing oxfendazole, such as Eqvalan and Systamex, are active against adult worms and larvae. Most other preparations, when given at normal dose rates, are active against adult forms only.

11 Weaning and Castration

Weaning

Six months or so after the birth of your foal, the day will come when he must be weaned from his dam. Most first-time horse breeders dread the prospect of the distress caused to both mare and foal at this time and put off the inevitable for as long as possible.

So why not just leave mare and foal together and let nature take its course? A mare in the wild, who is feeding herself and one foal, whilst carrying another right through the winter and early spring, when natural forage is sparse, is subject to an onerous burden. If she falls into very poor condition, the development and viability of the new pregnancy will be affected, but the suckling foal will not usually be weaned until the birth of the new foal.

Most domestic mares are adequately fed and do not suffer such rigorous lives. However, any mare is likely to continue to suckle her foal for a year, or even more if there is no new foal, until the youngster finally loses interest. A mare who has been put back in foal should have her

Fig 101   Ponies running free on Dartmoor in late summer.
The foal, unless sold in the autumn, will probably
continue to suckle until a new foal is born the following spring.

current youngster weaned at six months old to give her the best chance of maintaining her own condition and carrying the new pregnancy successfully to term.

Another reason for weaning your foal is to bring your mare back into work – it is often impractical to combine work with rearing a foal, although careful organisation can make it possible. Foals can be weaned as early as four months of age, if this is absolutely essential – for example, if the mare is ill – but it is advisable to wait until they are at least six months old, when they will be stronger and their bodily systems better developed. A foal should not be weaned unless he is healthy and accustomed to hard feed.

The trauma of weaning is exacerbated for both mare and foal if they are anywhere within earshot of each other after separation, so make arrangements to take your mare to a far field – even if this means boarding her with a friend or putting her at livery for a few weeks. If mare and foal cannot hear each other calling, they will settle down and accept the separation quite quickly.

The best place for a newly weaned foal is a large, safe loosebox, which should be made ready for him in advance. If you have one which has some natural light, e.g. through a roof light, or a window which is safely protected by steel mesh (but *not* unprotected glass), then this is a little less frightening for the foal when the door is fastened than being left in the dark.

The larger the loosebox, the better, as the foal will have room to get some exercise whilst he is shut in. A good thick bed should be laid down and there must be nothing projecting from the walls or left lying on the floor, which the foal could knock into or trip over if he dashes round the box looking for an escape back to mother. Do not use a haynet in a foal's box, but feed his hay in a manger, or, if necessary, once he has settled down enough not to charge all over it, on the floor. The water bucket is safest off the floor, in a holder, and feed should be given in a manger, not a container placed on the floor which can be turned up or tripped over and possibly cause an injury. There must be no gaps under the walls or doors, where a foal could catch a vulnerable foot or leg.

If these precautions seem excessively fussy, remember that a foal has no comprehension of his limitations and, when frightened or excited, is liable to rush blindly around in his panic, oblivious to danger. Serious accidents which could have been prevented with a little forethought, can happen in seconds.

Wean your foal in the earlier part of the day, so that you can keep an eye on him afterwards in daylight. The simplest method is to lead both mare and foal quietly into the loosebox, then take the mare straight out, leaving the foal behind. Fasten both top and bottom doors, or the foal may try to jump out. Take the mare off to her new field or temporary home immediately. If she has to travel, put on her travelling equipment before separating her from the foal, so that she can be led straight into the lorry or trailer, with the minimum of opportunity to think about what is happening.

If you know that the mare is possessive of her foal, you may need help to encourage her to move along. Equally, if the foal is well handled and used to human company, he may settle more quickly if someone stays to reassure him. A little food can often serve as a distraction.

If possible, your mare should have some quiet company when she is turned out, which will help her to stop fretting for her foal. Other mares are ideal, but an amenable gelding will do equally well. The mare should be turned into a fairly bare field or paddock where she can get plenty of walking exercise, but where a restricted nutritional intake will encourage her milk to dry up more quickly. This will probably take several weeks and a close watch must be kept on her for the first few days for any signs of mastitis.

Large studs use a different method of weaning foals. When the group is turned out for the day, one mare is taken out at first and turned out in a field some distance away, with other equine company. Her foal remains with the other youngsters and their mothers in his familiar environment and quickly adapts to the loss of his dam. Over a period of time, the other mares are removed, the final mare being left for as long as possible, to act as a restraining influence on the boisterous youngsters.

However, most private horse owners breed only one foal at a time, so it is unlikely that your foal will have any equine company of his own age when he is first weaned, unless you make arrangements with a friend who also has a lone foal. The advantage of finding a companion his own age for your weanling is that he will not suffer the loneliness of isolation, which is contrary to the nature of horses as herd animals. Two foals living together will quickly form a

Fig 102 *Thoroughbred weanlings turned out together in a large paddock with safe fencing. All are wearing leather head collars.*

mutual attachment and you may have difficulties when it comes to separating them later, but this always tends to be a problem where no more than two or three horses are sharing the same home. There are some horses who become very attached to a particular companion regardless of the presence of others, but usually separation from a larger group causes far less trouble.

A companion for your weaned foal might well be a good idea if you are unable to spend much time with him yourself, although you might want to wait for a day or two to see how he settles down. Only you can decide how the advantages or disadvantages might affect your management. Keep a close eye on your foal for the first day after his mother is taken away. He will appreciate your company in any case and this is the ideal time to increase his contact and familiarisation with humans. He will still have a strong urge to follow – his mother or anything that might replace her – and this can be put to good use in dispelling any nervousness he might have of people and in continuing to teach him to lead, without the encouragement of his dam moving alongside or ahead of him. Foals quickly learn to recognise the authority of their dams and of older horses. If you handle your foal with tact, sympathy and firmness, he will soon transfer that recognition to you and regard you as his 'herd leader'.

The safest course is to keep your newly weaned foal in for two or three days, before turning him out for the day in a safe paddock. Obviously, it is absolutely essential that the fences should be sound and secure and, if possible, use a paddock or field with which your foal is already familiar. When your foal is first turned out he is likely to gallop around the field, calling for his mother or for other horses, so keep watch until he has settled down to nibble at the grass. Check on him several times during the day.

Most breeders prefer to keep foals in at night from weaning until the following spring and this has the useful effect of continuing the foal's education – being led in and out and taught stable manners, as well as preventing the risk of him catching a cold or a chill in wet weather. However, it is not essential provided your youngster is well fed and some shelter, such as a thick hedge to windward or a field shelter, is available.

Do not change your foal's diet, but continue to offer him the same feed as when he was with his mother, adjusting the quantity according to his condition and appetite. He may lose a little condition immediately after weaning, but should soon pick up again as he gets used to this new phase of his life. Provided your feed is of good quality, with sufficient grazing, extra supplements are unnecessary, although milk powder may be a useful addition to the diet of a foal who has had to be weaned early. It is important to maintain a regular worming programme, especially in the spring and early summer.

There is no need to groom your youngster, except gently as a means of handling him. If he is living mostly out of doors, he should be left with the natural protection that develops in his coat.

Castration

Castration is defined as the surgical removal of both testicles from the male horse. We castrate for two reasons –

firstly to render the colt infertile, and secondly to remove the socially unwanted aspects of male behaviour. The operation can be performed at any age, but it is most usual to castrate a colt when he is between one and three years. The traditional time to operate is either during the spring or the autumn. These periods avoid the risk of post-operative infection from dust and flies.

The operation can be carried out in the standing position, under local anaesthesia with the colt sedated. The new sedative preparations are so powerful that this method is enjoying a resurgence of popularity as the safety of both surgeon and patient is more certain and the relaxed colt ensures that the operation can be carried out with few complications. However, the unexpected can always happen and a general anaesthetic allows a greater margin for error. For this reason, most veterinary surgeons operate on the recumbent colt under general anaesthesia when the operation can be carried out calmly, with good visibility and plenty of time to ensure that all the testicular tissue is removed.

There are many different techniques that can be used to perform the actual operation, but the end result is that the testicular tissue is removed and over the next few weeks the influence of the male hormones wanes and the colt loses his masculine behaviour. An important point to remember is that the colt is still able to cover a mare and get her in foal for a few weeks after the operation as there could be viable sperm still present in the spermatic cord.

Fig 103 Series of pictures showing a typical castration.
(i) The induction of the anaesthetic.

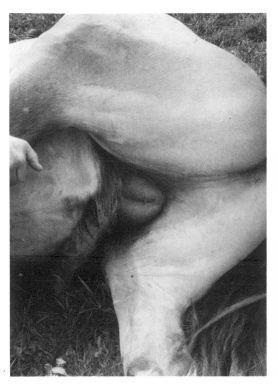

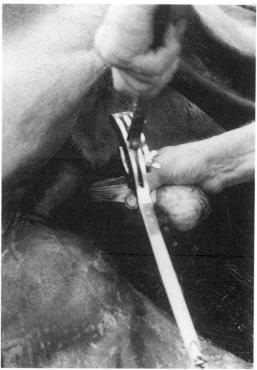

(ii)   Preparation of the operation site.

(iii)   Removing the testicle.

RIGS (CRYPTORCHIDS)

Occasionally, one or both testicles cannot be found in the inguinal region. In such cases the testicles are present but have not descended through the inguinal ring into the scrotum. These colts are called rigs (cryptorchids). They pose a problem, as if the visible testicle is removed, the testicle that is hidden in the ring or in the abdomen is still able to secrete male hormones and a horse that appears to be a gelding acts like a stallion.

To cope with these cases, the veterinary surgeon first explores the ring from the outside in an attempt to find the testicle. If the testicle is found then it can be gently exteriorised and removed; if not, then a major abdominal operation is necessary to find and remove it.

Some horses display stallion characteristics even though no testicular tissue can be found in the body and no male hormones can be identified in the blood. A laboratory test can be carried out which, by measuring testosterone levels, differentiates these 'false rigs' from those geldings in whom, due to faulty surgical technique, some testicular tissue has been left behind. No reason for this behaviour has been found; but it appears most commonly when a strange horse is introduced into a field of mares. This anti-social behaviour usually dies down after a few months.

153

12 Handling Youngstock

Summer passes, your foal is weaned and growing, but there are still three years before he can be backed. What do you do with him until then? A healthy young horse is lively and full of energy. Colts in particular may become boisterous and unruly and need firm handling.

For the first four to five years of his life, the young horse grows rapidly, gaining weight and strength. At six months old, the majority of foals are a physical match for most humans, so obedience must be learned from the start, otherwise you will never be sure of being in control – a potentially dangerous situation.

Obedience Training

Patience, calm and understanding are the tools required to teach a horse obedience;

Fig 104 *The first year's growth and development.*

Fig 105 The chestnut Welsh Mountain pony foal wants to play. He approaches his companion who stands his ground discouragingly.

not the whip. A horse who has been cowed into submission by the use of the whip has not learned obedience, but fear and resentment and he will rebel as soon as an opportunity occurs. The whip is used to educate and occasionally to punish, but it must never be raised in anger, however frustrating the circumstances.

Before he will obey you, your youngster needs to trust you and this is where developing your understanding of natural equine behaviour will help considerably. The horse has evolved to lead a social existence, for the mutual benefit of his whole herd. A dominant stallion leads and protects his herd and reproduces his own characteristics (which give him the potential to be dominant over other stallions) in his foals, thus giving them the best chance of survival and a successful

Fig 106 Biting at each other's hocks is a move used by adult fighting stallions.

Fig 107   Persistence is not yet paying off ...

*Fig 108   ... but the foal is too exuberant to give up
and will get a more active reaction in the end.*

life. An older mare will act as the stallion's deputy, with the other mares ranking under her according to their age, strength and temperament. Within this structure, the foals are protected, disciplined and educated. They quickly learn their place in the herd – at first subservient to everyone else – and to instantly obey the commands of their elders, which, in the wild, might be a matter of life or death. Soon, the peer group itself will have its own hierarchy, with the bigger, stronger foals dominating the smaller, younger ones.

For the horse, the need to co-operate within his herd is vital to survival and it is this willingness to adapt, coupled with the acceptance of a society based on differing levels of dominance of individuals which qualifies the horse so well for domestication and the demands made upon him by man.

Most domestic horses have to live either individually or in small peer groups, often in a highly artificial environment. Once your foal is weaned, you may be the only 'herd', or at least the only dominant associate, that he has left. If you have cared for him from birth, he will already be used to your presence and to obeying your simple commands – to lead, or to pick up his feet – and he will

Fig 109   Before any training can commence, trust must be established, and this is much easier when a foal has been handled from birth.

157

associate your appearance with the provision of food. In other words, your foal
will already have placed you in his social
hierarchy as a dominant 'herd leader'.
Your aim should be to maintain this
position throughout his training, whilst
expecting that, at times, as the youngster
grows up, your authority will be challenged.

It is much more difficult to establish
this basic relationship with a foal who has
never been handled before weaning. Of
course, he has to learn who is in charge,
but first he must be persuaded that this
strange two-legged creature can be trus-
ted and accepted not only into his social
group, but into his personal space. Trust
dispels fear and only then can education
begin, but a youngster who has not been
handled from birth is likely to challenge
your authority far more often in the early
stages of his life than one who is already
accustomed to your domination. (Dominance is always established by means of a
challenge, followed by submission or a
counter-challenge.) In your dealings
with your young horse, you must always
win these encounters, but without resor-
ting to the use of force or violence. Only
in the rarest circumstances, for example,
two stallions fighting over a group of
mares, will horses themselves indulge in a
serious display of force. At all other
times, a warning signal is enough for a
dominant animal to establish authority.
A similar admonishment should be
enough for the human horsemaster.

Foals turned out together play games –
chasing one another, boxing and staging
mock battles. These may be partly for
exercise, to test developing muscles and
to dispel surplus energy and high spirits,
but they are also one way in which a
group, close in age, sorts out its own

hierarchy. If your foal has no other
company except his mother (who will
soon get bored with indulging his youth-
ful exuberance), he may well do his best
to get you to join in the fun. Do not be
tempted, or you will have trouble later
on. Playing tag or shadow boxing might
seem like fun with a young foal, but it is a
different matter when he becomes a well-
grown yearling who has not learned
respect for humans.

If you have made a good start to your
foal's education during the first few
months of his life, the primary aim for the
next two to three years will be to consoli-
date this early learning, and gradually
introduce him to new sights and experiences. This should continue right up to the
point where he is ready to be backed,
towards the end of his fourth summer. It
is often tempting to back a young horse
considerably earlier than this, but young,
growing limbs and joints are soft and
easily damaged and giving the youngster
time to mature before starting to work
him will prolong his working life expec-
tancy by many years. Thoroughbred
racehorses are backed in their second year
and raced as two-year-olds, but the toll
taken in injuries is substantial.

Immediately after he is weaned is a
good time to establish your own relationship with your foal. He will be missing
his mother's reassuring presence and, if
he has no other equine company, will be
lonely. He can live out all the time, or in
at night and out during the day, once he
has settled down, but whatever his routine, try to make his surroundings interesting. For example, being stabled in a
yard where there are other animals – even
if they are not horses – will provide
company and interest. Human comings
and goings will also help. At the other

Fig 110   When introducing your youngster to different sights and sounds, do not forget other animals. It is surprising how many horses are nervous of sheep and cattle.

extreme, complete isolation, in a field or stable where he cannot see anything interesting going on around him, is likely to result in your foal becoming nervous, and afraid of everything he encounters when he eventually goes out into the world.

By now, your foal should be halter broken, i.e. used to wearing a head collar and to being led wherever you want him to go. If he is living out, catching him up regularly and taking him for a short walk, even if it is only back to his stable for a visit from the farrier, will remind him of what he has learned.

The foal should have his feet expertly trimmed from an early age, whenever the farrier comes to attend to the mare, which

should be at least every six weeks. If this procedure is begun when the foal is two or three months old, he will quickly learn to accept it and there will be no problem later when he needs to be shod. Prepare him for the farrier's first visit by teaching him to pick up his feet on command, in the same way as you would pick up an older horse's feet. Stand by the foal's near shoulder, run your left hand down the back of his foreleg and, leaning slightly against his shoulder, say, 'Pick it up', quietly and firmly, at the same time lifting the foot. At first, the foal will probably try to break away from your grip. If he does not give in easily, let go and repeat the exercise until he stands quietly. Avoid getting into a battle. Re-

159

peat the procedure for the other three legs, taking care to avoid being kicked when you come to the hind legs. A foal can lash out with lightning speed and can inflict a painful bruise even if more serious damage is unlikely. This is something to be aware of, whenever you are dealing with foals or youngstock.

Stable manners should be taught and expected from the beginning. You do not want to be knocked over by a growing youngster before you have a chance to empty his feed into the manger. He must learn to respect your personal space in the same way that he would that of a more dominant horse, so he should be taught to stand back when asked. This can be done by pushing his chest backwards, at the same time as giving the command firmly with your voice. If he does not understand or pushes past you, put the feed bucket down outside the door and spend a few moments on the lesson. Use one hand on his chest and the other above his nose and push him back, repeating the command several times, until he obeys. Reward him as soon as he takes a backward step. He will undoubtedly move forward again as soon as you retrieve the bucket, but repeat the command and immediately he moves one step back, empty the bucket quickly into the manger, so he can be rewarded with his dinner. Reinforcing the lesson each day will soon teach him to stand back until his feed is ready, but do not let him get away with shoving you unceremoniously aside. With an older youngster, whose education has been lacking, it may be necessary to show him a stick and perhaps tap his chest with it.

A young horse at grass does not really need much grooming, but brushing the mud off will accustom him to having his body handled and he can have his feet picked out. This is also a good time to teach him to move over, from one side to the other, in the stable. Standing at his shoulder, press your hand at the spot just behind the girth where you would give a leg aid and say, 'Move over'. To begin with, he will probably either scuttle sideways at speed, or ignore you completely. If he does not move, press the same spot again, a little harder this time, and repeat the command. He will soon get the idea.

A foal who has been taught to lead alongside his mother will readily follow you when he has been weaned and will soon learn to move up beside you, so that you can lead correctly, moving at his shoulder. A youngster who has not been taught to lead from birth must learn before he gets much older. If he has not been taught to lead, he is unlikely to be used to a head collar and he will probably have had little handling of any kind. The best time to start with these youngsters is immediately after weaning and before turning out, if you have bred the foal, or as soon as he arrives, if you bought him as a weanling.

If he is not used to a head collar, be prepared to spend time in the stable getting him used to the idea without frightening him. It may take some time to persuade him even to let you touch him and a week or more before he shows any signs of confidence. Patience is essential, but you must also be firm and determined to succeed. Eventually, the youngster will realise that you are not going to harm him and will look forward to your visits, which he will also associate with the provision of food. Progress will be made more quickly in this situation if the young horse is kept stabled. If possible, he should be turned into an enclosed yard or small paddock for an hour or two each day for exercise.

LEADING

To teach him to lead, a whip should be carried in your left hand, with the slack of the lead rope. Urge the young horse forward, saying, 'Walk on', in a firm and encouraging voice. If he balks, or pulls back, as he undoubtedly will, reach behind with your left hand and tap him on the flank with the whip. Be prepared for a sudden leap forward. Go with him and encourage him forward again. It will be easier to teach the youngster to lead in an enclosed space, such as a schooling arena, a yard or even a quiet narrow lane or driveway (with a closed gate at the end to prevent escape should he get away from you). Putting the horse between yourself and a fence can also be helpful in keeping him straight and encouraging him to move forward.

Most horses learn to accept the lead rein very quickly, and, with a little practice, will soon move out freely beside you on a loose rein. However, it is always advisable to carry a whip when leading young horses, in case of problems, when encountering strange objects or other horses, for example.

Tying up can be taught once the

Fig 111 The right way to lead a young horse. The long whip can be used to tap his flanks if he baulks or tries to swing round.

young horse is quiet, obedient and confident in the stable. As with everything else, the procedure should be carried out thoughtfully and safely. The old-fashioned practice of tying a youngster up and leaving him to get on with it is extremely dangerous for the horse. Not only will he be extremely frightened, but if he begins to fight the tie rope and throw himself around, the result can easily be fatal.

The horse should be wearing a well fitting head collar, preferably of leather, with the lead rope attached in the usual way, under the chin. Pass the end of the lead rope through the tie ring, but do not tie it. Hold the end of the lead rope whilst you groom the horse, keeping your actions calm and matter-of-fact. When the horse moves sufficiently to feel the tension taken up on the lead rope he will probably jerk back in surprise and anxiety. Give a little on the rope and speak soothingly to the horse until he relaxes. Continue grooming. After several repeats of this procedure the horse will learn not to argue with the rope and, in a few days, the rope can be tied with a quick release knot. A thick rope can be undone more easily than a thin one. Remember to tie up to a breakable cord attached to the tie ring and not to the ring itself. Then, if at any time the horse does become frightened and jerk back really hard, the cord will snap rather than his expensive leather head collar. A nylon head collar will not give in any case, so a safe breaking point is all the more essential if you use one of these. Never leave a young horse – or any horse for that matter – tied up unattended. You never know when he may take fright at something and panic.

The traditional way of handling horses is to do almost everything from the left (or near) side. However, this is thought to be a contributory factor in developing the 'one-sidedness' which is apparent in almost all riding horses, until they are schooled to be equally balanced on both sides. Some horses may even become shy or nervous of anyone approaching them from the right. It is, therefore, a good idea to handle your youngster equally from both sides, especially when leading him. In this way, he will learn to bend to both sides and not to favour only the left.

THE FOAL'S NEW ENVIRONMENT

Your youngster will encounter many different sights and sounds during his life. Whilst very young, like a small child, he has no fear. Soon, however, he will learn to treat everything unknown as potentially dangerous, until it is proved otherwise. The more he sees and becomes accustomed to whilst he is growing up, the less trouble you will have later, when his formal training begins. Even on a quiet hack through the countryside, many strange things can be encountered – cars, bicycles, tractors, motor bikes, dogs, dustbins, flapping polythene sacks, white lines across the road, water, pigs, ducks, umbrellas, the local hunt or pheasant shoot, road works, hedge trimmers, horse drawn vehicles. The list is endless.

Your horse needs to learn that if you tell him something is all right, he can trust you and whatever it is will not harm him. Most 'hazards' fall into obvious categories, for example, large threatening objects, moving objects, loud noises, strange smells, or a combination of any of these. Some hazards can be introduced to the horse, at home, gradually, taking care not to actually frighten him.

He may get used to the comings and goings of cars and other traffic around your yard, or by turning him out in a field alongside a road. Showgrounds are wonderful places for colourful sights and exciting sounds and, in this respect, showing your youngster, or simply taking him along when you are showing another horse, is an excellent addition to his education. Usually, there will be so much happening at once that he will not know where to look first. Of course, you should start with a small local show, before moving on to the busier large ones.

TITBITS

When, if ever, should your young horse be given titbits? The only real uses for titbits are as a distraction from something which is worrying the horse, or to encourage him to do something. For example, if he has suddenly developed an aversion to being caught in his field, a handful of nuts or a couple of slices of apple will often quickly change his mind. (Always slice apples and carrots lengthways to prevent them becoming stuck in the horse's gullet.) Horses who do not want to be caught are adept at snatching the proffered enticement and darting away before the head collar is on, so slip the lead rope round his neck whilst he is sniffing at your hand, *before* he gets the titbit.

Titbits are also useful if you have a horse who is unaccustomed to loading. A bucket with a few nuts shaken in front of him will often distract him from his fear

Fig 112   Showgrounds are wonderful places for exciting sights and sounds.

or reluctance. A horse who is nervous about something will often snatch a few mouthfuls of food whilst waiting to see what happens or whilst trying to decide how to react. This is known as 'displacement activity' and it is in this way that titbits can be used whilst training a young horse. For example, the horse might be nervous of being shod, or vaccinated, or having the saddle put on for the first time. A taste of his favourite treat at the critical moment will help to relax him and to prevent a possible resistance or explosion.

However, it is generally *not* a good idea to use titbits as rewards for good behaviour. When you want to reward the horse – which should be very frequently whilst he is being trained – stroke his neck and give him a word of praise. If you hand out titbits indiscriminately, especially to a young horse, he will constantly be looking for them. At best, he will make a nuisance of himself by continually nudging at your pockets and, at worst, develop a bad mannered inclination to lay back his ears and bite whenever the expected treat does not appear. The same rule applies when you visit a youngster in his field – a pat and a few words are a far better greeting than a titbit. An *occasional* treat does no harm, of course, and will encourage a shy youngster to come up to you, but do not let offering sugar lumps, mints or apples become a habit.

In all your dealings with your young horse, your attitude should be calm, firm and determined, but patient. A matter-of-fact approach is the most successful. Expect a horse to behave or obey and he usually will; expect him to be difficult and he will probably respond to your anticipation. If you are nervous or uncertain, he will wonder what you

are worried about and decide that he ought to be worried too.

Whenever you set out to teach a new lesson, make sure that everything is well organised beforehand and that you have all the equipment you need to hand, plus a helper if necessary. Think through what you want to achieve and know exactly how you are going to do it. Always use exactly the same tone of voice when giving a particular command. The actual words probably do not matter much, but the horse is extremely sensitive to tones of voice, which makes the voice an important aid in training. Use a lively tone to encourage him forward, a gentle one to reassure him, a warm tone for praise and a strong, sharp one for reprimands.

Young horses may need to be taught not to kick, bite or barge and any tendency to indulge in these habits must be nipped in the bud. Kicking is most likely if the horse is startled from behind, so avoid this and get him used to your presence at his rear when handling him. Let him know when you are going to move behind him by running your hand along his back. You are less likely to be kicked if you stand in close to the horse, rather than a few feet away. Until he is thoroughly used to you moving around him, be aware that he *might* kick and be ready to move out of the way. If he does lay back his ears and lash out, give him a hard slap and a loud reprimand immediately, then move in close again, letting him know that this behaviour will not be tolerated and will not prevent you from grooming his back end, or whatever you were trying to do.

Colts are more likely to take to biting than fillies, but both may go through phases of trying it. Again, it must be stopped immediately, with a smack and a reprimand, but do not make the mistake of hitting the horse on his sensitive nose. This is more likely to make him headshy than to stop him from biting. Your voice is the best deterrent, reinforced by a slap on the side of his neck.

Young horses are often headstrong and may attempt to barge out of their stables when the door is opened, or tread all over you and drag the lead rope from your hands when being led. This must be firmly discouraged. If necessary, hold up a stick in front of the horse, as a warning, when you go into his box, until he understands that he is not to make a beeline for the door. Be particularly decisive and determined in all your movements. Make him stand still whilst the head collar is fastened and carry your stick when you lead him out. Do not let him move off before you ask him to go and use your voice all the time to keep his attention. As colts become older and stronger, they can be mouthed and led out in a bridle instead of a head collar, which will give you more control.

Whilst it is important to teach a young horse good manners and to be firm, kindness and patience are also needed if the horse is to trust you and you are to understand each other. Do not assume that almost every move the horse makes is the start of an intention to misbehave. He is a living creature, not a machine and he may simply be making a gesture of companionship or affection.

At any time from weaning onwards, depending upon the size and strength of your youngster and what you plan to do with him, he can be mouthed, i.e. learn to wear a bridle with a bit. This will give you a better means of control when leading him out, will prepare him for

wearing a bridle later when ridden training begins and is necessary if you plan to show in hand.

Ideally, a mouthing bit, with keys, should be used. The keys encourage the youngster to play with the bit and to salivate. The bridle must be carefully fitted at first, so that the young horse will not be frightened. Unbuckle the off-side cheekpiece from the bit and the nearside cheekpiece from the headpiece. Remove the noseband, which will not be needed. If the horse is good about having his head and ears touched, slip the headpiece and browband over his ears. If he is nervous, slip the nearside of the headpiece out of the browband and put the headpiece over his poll, then quietly bring the browband round and slip the headpiece through. Do up the throatlash, leaving plenty of space between it and the cheekbone. Next, if the horse is standing quietly, buckle the bit to the offside cheekpiece. The important point is to avoid banging the horse's face or teeth with the bit if he starts to shake his head around. Bring the bit round and open the horse's mouth by inserting your right thumb between the bars in the usual way. Slip the bit in and buckle the cheekpiece to the headpiece.

To begin with, leave the bit in for about ten minutes. See that the horse is in a safe place, where the bit cannot get caught up on anything such as a door bolt. Gradually increase the period for which the horse wears the bit each day, until he is quite happy with it for an hour or so. For leading him out in a bridle, a mullen mouthed snaffle should be used – the flexible rubber type is ideal. The lead rein should either be attached to the off-side ring and passed through the nearside ring (when leading from the near side), or attached to a three-way coupling.

Fig 113   A mouthing bit with keys.

Some further work towards your horse's basic training can be achieved in his second or third year. If he has been shown, you will already have accustomed him to wearing rugs and travelling gear. If not, you can start by getting him used to a surcingle in the stable. Tighten it gradually, but do not leave him with it fastened too loosely, or it will irritate him or slip back and frighten him. Eventually, a saddle can be introduced.

Lungeing is not advisable until the horse is at least three years old, as the risk of straining his joints is considerable when working on a circle. However, long reining is invaluable for building up a young horse's muscles and teaching him to go forward and accept the bit. It is well worthwhile learning how to long rein correctly, provided you have a safe place to work, such as in private lanes or tracks.

13 Owning a Stallion

For many years the popular view has been that the entire horse has no place in competitive sport, mainly for reasons of safety and the inconvenience of having to take special precautions in handling him. His strength and power, his unpredictability and the potential difficulties of coping with him in the vicinity of mares likely to be in season, all seemed good reasons for not using him as a riding animal. However, the demand for 'proven performance ability' in breeding stock has begun to reverse this opinion and the number of competing stallions in all four competitive disciplines is gradually increasing.

Of course, a stallion does need firm, knowledgeable and carefully considered handling and it would be unwise to contemplate keeping one without some previous experience. It is essential to be confident about your ability to cope, both in the saddle and on the ground, since a stallion will soon take advantage of any lapse in your authority. Suitable facilities for keeping a stallion are also essential. However, provided he learns early on that humans must be respected,

Fig 114   Welton Apollo, the successful Thoroughbred eventing stallion, competing at Badminton, ridden by Leslie Law.

*Fig 115   A well-managed stallion makes
a superb saddle horse. Andrew Lorent,
at the age of nine, performing half
pass on the Anglo-Arab, Le Maréchal.*

there is no reason why he cannot be ridden and worked like any other horse and pursue a successful dual career.

What are the advantages of owning and riding a stallion? Obviously, a stallion has much more natural 'presence' than a mare or gelding and a competent rider can make the most of this in showing or dressage classes. Also, if you enjoy a lively, spirited ride, a stallion will give you this challenge. Thirdly, a stallion is invariably powerful, onward going and courageous, which can be a considerable advantage in eventing or show-jumping.

Lastly, a good stallion can help earn his own keep doing duty as a stud horse.

For the working horse owner and rider, the major disadvantage of owning a stallion is that he cannot simply be turned away with other horses if you are away from home or unable to ride him for some reason. Special arrangements need to be made for his safe management. Many livery yards also refuse to take stallions.

Many people come to stallion ownership through buying or breeding a colt foal and keeping him entire. This is perhaps the best way to begin, since you will have the advantage of knowing the animal from birth and the assurance that he has not been neglected or incompetently treated. He can be taught to respect your authority from the start and you will be accustomed to handling him on a daily basis as he grows up. Unlike mares and geldings, a stallion may become more unpredictable and difficult to handle as he ages beyond his prime, so buying an older stallion is probably not a good idea for the first-time stallion owner.

The particular needs of the stallion and the best method of management depend upon his breed and what you want to do with him. Thoroughbred stallions, being large, powerful and high spirited, as well as often extremely valuable, are usually subject to a more artificial life-style than other breeds. Few are ridden and their exercise tends to be limited to the minimum sufficient to keep them fit both physically and mentally. Safe, secure premises are necessary and keeping a Thoroughbred stallion is a job for the professional, beyond the scope of the private horse owner whose horses are a hobby rather than a full-time occupation.

Warmbloods have a calmer temperament and are often kept for show-

*Fig 116   Cathy Brown with her top endurance riding horse,
the pure-bred Arab stallion, Maquib.*

jumping and dressage, whilst stallions of smaller breeds may be very successfully kept as riding horses. Two of the most popular are the Arabian, kept for showing and long-distance riding but also an excellent all round performer, and the Welsh Cob, used for showing, riding, driving and often crossed with the Thoroughbred to breed competition horses.

Handling Entires

The young entire should be handled in much the same way as any other young horse, but with special regard to discipline. The best companion for a young colt is another colt. Colts are lively, demanding young animals who will play and have mock battles whilst growing up together, but they will tend to annoy other horses. Once they are about a year old, they will attempt to mount mares and fillies, although they will not be fertile and able to sire foals until they are two years old or more.

It is important to handle young entires as much as possible and to develop a relationship based on trust and respect. The more time you give to your colt, the easier he will be to handle later. Showing in hand will help educate him and get him used to taking the presence of other strange horses for granted.

As far as possible, treat your colt like

169

Fig 117   Early discipline is essential to control the dominant attitude of the male horse. This young Welsh Cob shows commanding presence and powerful conformation.

any other horse, whilst remaining mindful that he *is* a stallion. Expect good behaviour, not explosions. Care must obviously be taken when out in company not to upset other horses – do not get too close and be sure that you have your colt under proper control. Always lead him in his bridle and carry a whip (or a cane, for showing).

Boxes

At home, the young stallion's box should be as large as possible, safe, with a strong door and secure fastenings. There should be a top door which can be closed whenever he is likely to become over-excited and attempt to jump out, for example if mares are being moved around. A framed wire grille, which can be closed for security but still allows the horse to see out, is also a useful addition to stallion boxes.

Stallions will live quite happily in boxes adjoining other horses and will appreciate the company. However, care must be taken if other horses are in the vicinity when your entire is turned out. In winter, there is usually little problem, but once the breeding season begins, he is likely to try jumping out to reach mares in another field, so your hedges or fences need to be high and solid enough to

Fig 118 An attractive two-year-old colt by the event horse, Welton Crackerjack (three-quarters Thoroughbred/ one-quarter Irish Draught), out of a seven-eighths Thoroughbred mare.

Fig 119 Stallions need to be turned out for exercise and relaxation, like any other horse. This Andalusian stallion is enjoying the freedom of an outdoor schooling arena.

Fig 120   A stallion who is going to be used for stud work needs to start the season in good condition. This pure-bred Arabian, Rahqui, is seventeen years old.

prevent escape. Also, entires turned out together may fight seriously for dominance if there are mares about in spring, and they may attack geldings.

Stud Work

Given the opportunity, a two-year-old colt may be capable of getting mares in foal, but it is advisable to wait until he is three and has had a chance to develop more maturity himself, before formally introducing him to stud duty. In his first season, he should be limited to a few mares and they should preferably be older, experienced ones, who will stand quietly for him, accept his advances and let him learn his job without frightening him.

A stallion who is going to be used for stud work must start the season in good condition, which should be maintained throughout the summer. He will need a high-energy diet including concentrates, regardless of whether he is also being ridden. The protein content should be comparable with that given to the pregnant mare and stud cubes are an easy way of ensuring that a correctly balanced ration is fed.

Hay must be of good quality and is best fed ad lib to stabled horses, whilst those running out should be offered hay until they disregard it in favour of new spring grass. Stallions who have to be stabled most of the time benefit greatly from cut green feed – either grass or, ideally, lucerne (alfalfa), which has a high protein content. This must be cut freshly each day and any left uneaten should be removed at the end of the day, before it wilts. Alfalfa nuts are also available.

A stallion with a full book of mares will need all of his energies for stud duties during the season, so remember that the number of mares you accept will have to be limited considerably if you also want to work your stallion under saddle. You will have to work out how best to combine your horse's stud work with his ridden work. For example, it might be more convenient to have him available for covering early in the season, before you start concentrating on riding and competitions, or to fit stud duties into a lull between competitive phases, or to cut down riding around mid-summer, letting him finish off the season covering mares. Of course, you can combine both in parallel, but the horse must not be overworked.

If your stallion has been well schooled, there should be no difficulty in combining riding with stud duties. Many stallion owners claim that their horses know exactly what to expect as soon as their tack is put on – a covering bridle for stud work and a working bridle and saddle for ridden work – and they behave accordingly. A stallion who is also worked under saddle and ridden out is likely to be much more relaxed, happy and easy to handle than one who is never ridden, just as a horse running out with his mares is more contented and amenable than one who is kept stabled most of the time. Also, your stallion's competitive

achievements will help promote his value as a stud horse.

If you accept mares to run out with your stallion, be sure that their hind shoes have been removed and that they have been wormed recently. All visiting mares should have up-to-date vaccination certificates and, ideally, should have been swabbed for contagious diseases. Before starting the season's stud work, your stallion should also have a check-up.

There is no longer a statutory requirement in Britain for stallions to be licensed. However, many breed societies run their own licensing schemes and whatever breed of horse you choose, you should ensure that you have complied with all the rules of the breed society concerned. Otherwise, you may find that your horse's offspring are unregisterable and, therefore, of greatly decreased value.

Before you decide to keep a young colt as an entire, there are various points you should consider. Firstly, is his breeding such that he would be likely to be in demand as a future sire? Are his blood lines recognised and is he registered, or capable of being registered with a breed society? This is seldom a problem with pure bred animals, most of whom are registered at the appropriate time after birth. However, a part-bred or cross-bred horse is a more doubtful proposition. There are some very successful part-bred stallions, but the cross has to be one which is of accepted usefulness and value.

The most obvious cross – which is, in fact, known as an independent breed – is that of the Thoroughbred with the Arab, which produces the Anglo-Arab. Other accepted crosses include the Thoroughbred with the Irish Draught, producing top-class competition horses, or the

Fig 121   Good conformation is essential in a breeding stallion. Kalisz, a two-year-old pure bred Arabian colt, is a lovely example of his breed.

Thoroughbred with Warmbloods, producing a lighter, more elegant type of Warmblood. A cross of the Arabian with the Hispanic breeds might also provide an attractive proposition.

Crosses lacking any direct 'hot blood' content are unlikely to be successful as breeding sires, since the characteristics of their offspring are too variable to predict with any degree of accuracy and their ability to reproduce themselves is questionable.

The next point to consider is the conformation of the colt concerned. Is he a good example of his breed, as well as possessing basically sound conformation? Is he sound in wind, limb and eye and is his action correct? Finally, what are the competitive or working achievements of his parents and grandparents? Are they sufficient to warrant further breeding from the line?

Many people breed to show, rather than to produce ridden competition horses, and there is obviously a great deal of satisfaction to be obtained from breeding and producing a show champion from your own mare and stallion. Pony stallions are undoubtedly easier to manage than their larger counterparts and since even the smaller breeds can also be successfully ridden, this may be the most attractive option for the private owner who cannot run a commercial stud but has an interest in breeding.

Before you embark on such a project however, remember that champions are few and far between and it is frequently professionalism in producing a show animal that reaps the major rewards, not simply the superiority of the particular exhibit.

In the final analysis, a horse or pony should only be bred because it is wanted for its own sake, with its future assured. For every successful, accidentally bred 'freak' in the horse world, there are many who live out sad lives without amounting to anything and it is *selective* breeding, with due attention to the welfare of the horse, which should be every horse breeder's aim.

14 Future Trends

It is rather exciting to realise that an aspect of the horse world as traditional and slow to change as breeding has such a challenging future. One aspect of this future that will have to be redressed is artificial breeding, i.e. the use of artificial insemination as a method of getting mares in foal and the practice of embryo transfer. The practical aspects of both these techniques have, in the large part, been solved. What is needed now is for the various breed societies to agree on a code of practice.

For the time being, if anyone wishes to try out either of these techniques, it is advisable to check first with the relevant breed society to find out what their current rulings are. For those who are interested, what follows is a brief update on both procedures.

Artificial Insemination

Artificial insemination can be divided into two procedures – the collection of semen from the stallion and the insemination, that is, the placement of the semen into the uterus of the mare. When done correctly, the conception rates achieved approach those obtained by natural service.

What are the advantages and disadvantages of this technique? The main advantage of AI is the effect it has on increased stallion utilisation. In the question of numbers of sperm per ejaculate, nature has been very generous and each ejaculate can be divided many times, each portion being used to get a mare in foal. It is possible that the fear of the overuse of popular sires and a further concentration of blood lines is one of the reasons why the authorities do not allow its use in the Thoroughbred industry. However, the advantage of being able to plan the stallion's book and reducing his work-load is a considerable advantage of AI.

The use of AI minimises the risk of venereal disease as the dilution of semen with extender, containing antibiotic preparations, reduces the number of such pathogens. Likewise, this treatment reduces the amount of uterine contamination and, therefore, increases the chances of successful conception in mares with a low level of uterine immunity.

Cooling techniques, where the temperature of the semen is slowly lowered to 5°C, allow the sperm to remain viable for up to forty-eight hours and deep freezing can be used to keep semen indefinitely. Unfortunately, resistance of the various breed societies to these techniques has restricted their development and, as yet, the viability of sperm which has been deep frozen is not good. An advantage of both these methods is the ability to store semen and transport it long distances or to delay insemination until a more convenient time, thus reducing the considerable costs of transporting the mare to the stud.

Once the practice of deep freezing sperm is accepted, long-term storage of scarce genetic material for future use is probably the most important application for AI. The danger of over-concentration of one blood line has already been mentioned, but another common fear of the technique – that of fraud – is, in my opinion, unfounded. It is no more likely that the wrong semen sample will be used to inseminate a mare than, as can happen at present, the wrong stallion gets to the mare. Also, if all breed societies accepted blood typing of foals, the question would not arise.

Attendants and all who are involved in handling the mare and stallion during a normal covering are exposed to risk and there is no doubt that the collection of semen from the stallion adds to that risk. It should be common practice for those in the firing-line to wear protective clothing when collecting semen.

The extra cost of AI is an obvious disadvantage. The personnel responsible for the collection of semen need to be trained and a veterinary surgeon will normally be involved in the preparation and assessment of the semen and the actual insemination of the mare. All this incurs expense. The procedure of AI can be divided into two main areas:

1. The collection of the semen. The most common way of collecting semen is to use an artificial vagina, which is a tube with a rigid outer casing and a flexible, removable inner lining. The space between is filled with water kept at blood temperature. Most stallions can be trained to ejaculate into this tube and the sperm-rich fraction is caught in a flask attached to the free end. If a quiet mare in oestrus is difficult to find then a suitable wooden mount, called a phantom mare, can be used. Most stallions will accept such a mount as a sexual object. This method is certainly safer for personnel and avoids the risk of injury to the mare.

Stallion condoms, either special ones or a plastic sleeve, can be used to collect semen, but they are dangerous to apply and the ejaculate is always more contaminated than a sample taken into an artificial vagina. Some stallions also resent their use and refuse to ejaculate into them.

Alternatively, a sample of semen can be removed from the mare's vagina or collected as the stallion withdraws. This post ejaculate sample is invariably contaminated, thus negating the minimal contamination ideal of AI, and the quantity and viability of the sample is often poor.

2. The treatment of the semen and insemination of the mare. Once the semen has been obtained, a sample is usually evaluated for sperm quality. The motility of the sperm is judged and the ratio of normal to damaged sperm and the density of the sample is calculated. If the sample is satisfactory, then a suitable expander is added and the insemination volume necessary to inseminate 500 million sperm is calculated.

If the semen is to be stored, it is now that the regulated cooling takes place. Spermatozoa are delicate organisms and must be treated with care if an acceptable number are to survive. It is especially important to avoid contact with spermicidal chemicals and to avoid exposure to light.

The actual insemination of the mare is a relatively easy matter. The insemination dose is administered with a syringe, again being sure that no spermicidal chemicals are in contact with the sample, through the cervix into the body of the uterus. The ideal time for insemination is just before ovulation, but if this cannot be determined then insemination every forty-eight hours starting forty-eight hours from the start of heat, usually produces results.

Embryo Transfer

Embryo transfer involves the transfer of an embryo, generally eight days old, from a donor mare to a recipient. The technique has been well proven and results have been good. The advantages of embryo transfer are varied. The main advantage is probably the ability to obtain six to eight embryos from one mare in each breeding season. These could all be implanted into recipient mares and, in theory, all the mares could foal down in the succeeding year. In practice, half that number of embryos actually 'take' but that still means three or four foals, all with selected parentage, born each year. This is a very important help for those people concerned with increasing the numbers of rare breeds such as the Przewalski horse and is of equal importance to those interested in increasing the numbers of good show or performance horses, since foals can be produced from competition mares without taking them out of work. Using this technique would also improve the chances of getting a foal from an old mare, who, for one reason or another, can no longer carry a foal.

The main disadvantage of embryo transfer is cost. At the moment it is only possible to obtain one embryo from each collection, unlike the situation in cattle where drug-induced superovulation means that the cow can produce many embryos. The long follicular stage and difficulty in timing ovulation in the mare, even using synchronising drugs, mean that a number of recipient mares must be kept, to ensure that one is suitable for implantation at the right time.

The methods employed to carry out embryo transplant are basically simple. The donor mare is monitored closely and when ovulation is deemed to have taken place, determined either by ultrasound or manual examination, the mare is inseminated. The first insemination is at forty-eight hours post ovulation, repeated every forty-eight hours until the end of heat. At day eight after ovulation, the embryo is washed out of the uterus by flushing the uterus with saline through a special catheter. The embryo is stored in a culture dish until the recipient mare is ready.

The recipient mare should be at the same stage of her cycle as the donor and the embryo is introduced into her uterus using a special pipette. Absolute cleanliness is essential to avoid the introduction of any infective agents. A surgical technique, where the embryo is introduced into the uterus via a flank incision has also been used.

The results that can be obtained from embryo transfer are quite good enough to make it a viable procedure. If both the donor and recipient mares are healthy and have normal cycles, then the success rates following surgical transfer should be above 70 per cent and following non-surgical transfer, between 50 and 60 per cent.

Appendix 1

Some Breed Societies and their Breeds

BRITISH NATIVE PONIES

The Shetland Pony Stud-Book Society

Address Pedigree House, 6 King's Place, Perth PH2 8AD.

Breed Description Height must not exceed 40in at three years nor 42in at four years and over. Any colour acceptable except spotted. The smallest of all breeds, with tremendous strength for size. Small refined head, small neat ears and large eyes. Strong neck, short well ribbed-up back, strong quarters and high set tail. Plenty of bone for size, short cannons. Plenty of silky, straight hair on mane and tail. Hard, round open feet.

Exmoor Pony Society

Address Glen Fern, Waddicombe, Dulverton, Somerset, TA22 9RY.

Breed Description Mares not to exceed 12.2hh; stallions not to exceed 12.3hh. Colours: bay, brown or dun, with mealy markings on muzzle, around eyes and inside flanks; no white markings permitted. Wide forehead, short thick ears, 'toad' eye, legs clean and short.

The Dartmoor Pony Society

Address Slade, Hexworthy, Princetown, Nr Yelverton, Devon.

Breed Description Height: not exceeding 12.2hh. Colour: bay brown, black, grey and sometimes roan or chestnut; no piebalds or skewbalds allowed and excessive white is to be discouraged, Head: small, well set on and 'blood-like'. Ears: very small and alert. Neck: strong, but not too heavy and neither long nor short – stallions, moderate crest. Shoulders: well laid back, giving a good front and the feel of a 'little horse' when ridden. Back, loins and hindquarters: strong and well covered with muscle. Feet: tough and well shaped. Action: low, free, typical hack or riding action. Tail: set high and full.

New Forest Pony Breeding and Cattle Society

Address Beacon Corner, Burley, Ringwood, Hants BH24 4EH.

Breed Description Maximum height 14.2hh. Any colour permissible except piebald, skewbald or blue-eyed cream. Bays and browns predominate. White markings permitted on head and legs. Pony of riding type, with substance. Well set on pony head, long, sloping shoulders, strong quarters, plenty of bone, good depth of body, straight limbs and hard, round feet. The larger ponies are capable of carrying adults.

Fell Pony Society

Address 19 Dragley Beck, Ulverston, Cumbria LA12 0HD.

Breed Description Maximum height 14hh (ideally, about 13.2hh). Permitted colours: black, brown, bay and grey. A star or a little white on a hind heel are allowed, but much white on face or legs indicates cross-breeding. The pony should be strong and active, lively and alert, with great bone. The head is small, with large nostrils, large, bright eyes and short ears. The shoulders are sloping, the body strong and deep, with well-muscled quarters. The legs must be strong, with plenty of flat bone and plenty of fine, silky feather. The feet must be round, with the characteristic blue horn. The stride should be long, with good knee and hock action.

The Dales Pony Society

Address 196 Springvale Road, Walkley, Sheffield S6 3NU.

Breed Description Maximum height 14.2hh. Colours predominantly black, with some brown, grey, bay and, rarely, roan. Renowned for hard, round, open, well-shaped feet and legs with dense, flat bone (average 8–9in) and ample silky, straight feather. Action straight and true, really using knees and hocks. The head should be neat, showing no dish, but broad between the eyes; muzzle rela-

tively small; incurving 'pony' ears. Long foretop, mane and tail of straight hair. Muscular neck of ample length, set into sloping shoulders. The body should be short coupled with strong loins and lengthy, powerful quarters.

The Highland Pony Society

Address Beechwood, Elie, Fife KY9 1DH.

Breed Description Height 13 to 14.2hh. Colours: various shades of dun (mouse, yellow, grey, cream, fox), also grey, brown, black and occasionally bay or liver chestnut with silver mane and tail. Most ponies have the dorsal eel stripe and many have zebra markings on the forelegs. A small star is permissible, but other white markings are discouraged. The head should be broad between alert and kindly eyes, short between the eyes and muzzle, with the nostrils wide. The neck should be strong, not short, with a good arched topline, the throat clean. The shoulder should be well sloped and the withers pronounced, with the body compact, the chest deep and the ribs well sprung. The quarters should be powerful, with well-developed thighs and second thighs. The legs should have flat hard bone, with strong forearms, broad knees, short cannons, oblique pasterns and well-shaped, dark hooves. Feather should be silky and not too heavy, ending in a prominent tuft at the fetlock. The mane and tail should be long, silky and flowing, with the tail set fairly high and carried gaily.

Connemara Pony Breeders Society

Address   73 Dalysfort Road, Salthill, Galway, Ireland.

English Connemara Pony Society

Address   2 The Leys, Salford, Chipping Norton, Oxon OX7 5FD.

Breed Description   Height 13 to 14hh. Colours: grey, black, bay, brown, dun, with occasional roans and chestnuts. The body should be compact, deep, standing on short legs and covering a lot of ground. The head and neck should be well balanced and the shoulders of riding type. The legs should have clean, hard, flat bone, measuring approximately seven to eight inches below the knee. Characteristics are hardiness of constitution, staying power, docility, intelligence and soundness.

The Welsh Pony and Cob Society

Address   6 Chalybeate Street, Aberystwyth, Dyfed SY23 1HS.

Breed Descriptions

1. *Section A*. Height not exceeding 12hh Any colour except piebald and skewbald. Head small, clean-cut, well set on and tapering to the muzzle; eyes bold; ears small and pointed, well up on the head, proportionately close; nostrils prominent and open. Jaws and throat should be clean and finely-cut with ample room at the angle of the jaw; the neck should be lengthy, well carried and moderately lean in mares but inclined to be cresty in mature stallions. Shoulders should be long and sloping well back, with the humerus upright so that the foreleg is not set in under the body. The forelegs should be set square, with long, strong forearms, well developed at the knee and short flat bone below the knee. The hind legs should have large, flat, clean hocks, with the hind leg not too bent. The feet must be well shaped and round, the hooves dense. The back and loins should be strong, muscular and well coupled, with the girth deep and hindquarters lengthy and fine. The tail should be well set on and carried gaily. The action is quick, free and straight from the shoulder, with the hocks well flexed, with straight, powerful leverage and well under the body.

2. *Section B*. Similar to Section A, except that the Section B pony should be a riding pony, with quality, riding action, adequate bone and substance, hardiness and constitution and with pony character.

3. *Sections C and D*. (Section C, Ponies of Cob type, not to exceed 13.2hh and Section D, Welsh Cob.) Strong, hardy and active, with pony character and as much substance as possible. Any colour except piebald and skewbald. The head should be full of quality (a coarse head and Roman nose are objectionable) with eyes bold, prominent and set widely apart, ears neat and well set, neck lengthy and well carried, moderately lean in mares, but inclined to be cresty in mature stallions. The shoulders should be strong, but well laid back, with the forelegs set square with long, strong forearms, well-developed knees and an abundance of bone below them. Pasterns of proportionate slope and length with well-shaped feet and dense hooves. When roughed off a moderate quantity of

silky feather is permissible.

The back and loins should be muscular, strong and well coupled. The animal should be deep through the head and well ribbed up. Hindquarters should be lengthy and strong with the tail well set on. Hind legs should have strong muscular second thighs, large, flat and clean hocks, with the hock not set behind a line falling from the point of the quarter to the fetlock joint. The action should be free, true and forcible. The knee should be bent and the whole foreleg should be extended straight from the shoulder and as far forward as possible in the trot. Hocks flexed under the body with straight and powerful leverage.

OTHER BREEDS

Irish Draught Horse Society of Great Britain

Address   4th Street, NAC, Stoneleigh, Kenilworth, Warwickshire.

Breed Description   At three years old, stallions should be 16hh and over, mares 15.2hh and over, with 9in or more of clean, flat bone. Any strong whole colour, including grey. An active, short shinned, powerful horse with substance and quality. Bold eyes, set well apart, wide forehead and long, well set ears. Head generous and pleasant, not coarse. Shoulders clean-cut and withers well defined with the neck set on high and carried proudly, showing a good length of rein. Forearms large and generous, set near the ground; cannon bones short and straight; pasterns strong and in proportion; hooves hard and sound. The back should be strong and the girth deep, with

strong loins and quarters, the croup long and gently sloping. The upper thighs should be strong and powerful, the second thighs well developed, the hocks sound and generous and in a line from buttocks to heels. The action should be smooth, straight and free, without exaggeration, but with good flexion of the hocks.

British Trakehner Association

Address   Buckwood, Fulmer, Buckinghamshire.

Breed Description   Striking, elegant presence, size and substance. Classical line breeding. A refined head, broad forehead, neatly formed muzzle, large kind wide-set eyes, clearly defined jawbone. Throat clear, set on a graceful neck, which, in turn, is set on the shoulder to provide balance. Legs straight, medium pasterns. Sloping shoulders to give freedom of movement at any gait. Powerful hocks giving the strength to perform intricate dressage movements and the power for jumping.

The Arab Horse Society

Address   Goddards Green, Cranbrook, Kent TN17 3LP.

Breed Description   No height limit, but generally between 14.2 and 15.2hh. Colours: grey, chestnut, bay, brown and black. Head extremely refined with clearly defined bone structure, dished profile desirable; eyes large and dark; nostrils large, expressive, flexible and capable of great expansion. Muzzle small, ears finely chiselled with tops often curved. Cheek-bones wide apart with plenty of throat room. The neck has an arched appearance and must be cleanly modelled and spring from the top of the chest. The shoulder is well laid back, long and clearly defined at the withers, the chest deep and reasonably wide. The back is short and strong with a slight concave line between withers and loins, which should spring strongly, in a curve, to the quarters.

The body is deep through and quite wide, the ribs being round and well sprung. The line of the quarters should be nearly horizontal, with the tail appearing as a natural extension to this line. The tail carriage is distinctively elevated, particularly when the horse is moving or excited. Forearm long and strongly muscled with large flat knees and short cannon bones; pasterns of reasonable length and slope with elastic action. Hocks large and clean, hind pasterns a little steeper than in front. Fore hooves circular and open; hind hooves more oval, with very hard horn. Coat fine and silky; skin fine, velvety and dark (pink under white markings); mane and tail fine and silky. Legs and heels clean.

The Lusitano Breed Society of Great Britain

Address Foxcroft, Bulstrode Lane, Felden, Hemel Hempstead, Herts HP3 0BP.

Breed Description Height generally between 15.1 and 16.2hh. Any true colour, including dun and chestnut. Long, noble head, with straight or slightly convex profile narrowing to a long, finely curved nose. Large, generous eyes, inclined to be almond-shaped. Long powerful neck, deep at the base and set at a wider angle to the shoulder than in the Thoroughbred, giving the impression of being more upright. A high wither, powerful shoulder and deep ribcage, which is slightly flat at the sides. A short-coupled body with broad, powerful loins and a gently sloping croup with the tail set low rather than high. The hind leg positioned well underneath the body axis, producing excellent hock action and powerful forward impulsion, which also helps easy collection. Fine legs with dense bone. Courageous, willing and gentle temperament.

The British Lipizzaner Horse Society

Address Ausdan Stud, Glynarthen, Llandysul, Dyfed SA44 6PB.

Breed Description Height ideally ranging from 15 to 16hh. Colour predominantly white although other colours do occur. The white coat is attained in degrees. Head often lean and large, with a 'broken' nose line; ears large and eyes intelligent. The neck is muscular and high set, the chest broad and deep, but not especially pronounced. The back is often long, but always strong and the loins long, rounded and muscular. The shoulders are often short and steep. The tail is set high and well carried. The legs tend to be short with the upper leg longer than the lower; medium length cannons and wide, well-defined joints, with short, elastic pasterns. The hooves are small and tough. Gait high stepping and energetic.

The British Quarter Horse Association Ltd

Address 4th Street, National Agricultural Centre, Stoneleigh, Kenilworth,

Warwickshire CV8 2LG.

Breed Description Height usually between 14.3 and 15.1hh, but can range from 14 to 16hh. Short, broad head, small 'fox ears', wide, kind eyes and large, sensitive nostrils. Short muzzle and firm mouth with well-developed jaws. The head joins the neck at a near 45 degree angle, the neck is of medium length, slightly arched and blending into a deeply sloping shoulder. The withers are sharp and extend well back, giving a good saddle position. Deep, broad chest with great girth and wide-set forelegs with powerfully muscled forearms, smooth joints and very short cannons set on clean fetlocks, medium-length pasterns and sound feet. Short back, close-coupled with powerful loins, the barrel formed by deep, well-sprung ribs. The hindquarters broad, deep and heavily muscled through the thigh, stifle, gaskin and down to the hock. Hocks wide, deep, straight and clean. The feet are well rounded with open heels. The action is collected and the Quarter Horse turns and stops with noticeable ease and balance, with his hocks always well under him.

The Hackney Horse Society

Address Clump Cottage, Chitterne, Warminster, Wiltshire BA12 0LL.

Breed Description Ponies up to 14hh, horses over 14hh. Usual colours: bay, dark brown, chestnut, black. Head small and convex, with large eyes, small ears and muzzle. Long, well-formed neck, powerful shoulders with low withers. Compact body with great depth of chest. Tail well set on quarters and carried high. Forelegs straight, gently sloping pastern to well-shaped feet. Strong hocks, well let down. Fine silky coat. Action must be fine, with the leg raised and thrown forward to cover the ground, straight and true. Brilliance and correctness are paramount with the head erect and ears pricked.

British Morgan Horse Society

Address George and Dragon Hall, Mary Place, London W11.

Breed Description Height usually between 14.1 and 15.2hh. Any colour but not white. The head should be expressive with broad forehead, large, prominent eyes, with straight or slightly dished short face, firm, fine lips, large nostrils and well-rounded jowl. Ears should be short, set rather wide apart and carried alertly. Throatlatch slightly deeper than other breeds and refined sufficiently to allow proper flexion at the poll. The neck should come out on top of a well-angulated shoulder and the Morgan is renowned for its generous crest in both sexes, which should be slightly arched and blend with the withers and back. The top line should be considerably longer than the bottom line. The withers should be well laid back and the body should be compact, with a short back, close coupling, broad loins, deep flank, well-sprung ribs, croup long and well muscled, with tail attached high and carried gracefully and straight. The legs should be straight and sound with short cannons and flat bone, the forearm proportionately long and the pasterns sufficiently long and angulated to give a light, springy step. The feet should be in proportion, round, open at the heel, with dense horn.

British Appaloosa Society

Address 2 Frederick Street, Rugby CV21 2EN.

Breed Description Height of 14.2hh and over. Colour: a distinguishing factor, with eight basic patterns:

1. *Spotted Blanket.* Dark forehand with white over loin and hips with round or egg-shaped spots.
2. *White Blanket.* Dark forehand with a blanket devoid or nearly void of spots.
3. *Marble (or roan).* Base colour usually red or blue roan, usually with 'varnish marks'.
4. *Leopard.* Base colour pure white with evenly distributed dark spots over entire body.
5. *Near Leopard.* Born with Leopard coloured body markings but with different coloured head and legs, or head, shoulders and legs.
6. *Few Spot Leopard.* Basic colour white with blue or red roan 'varnish marks' and occasional spots.
7. *Snowflake.* Base colour dark, with white spots over the body.
8. *Frosted Hip.* Dark base colour with either frost or white spots on loin and hips.

They are hardy, sure-footed, active and good 'doers', with tractable temperament. Head straight and lean, ears pointed and medium sized. Lips, muzzle, nostrils and around the eyes generally showing parti-coloured skin. Eyes dark, surrounded by white sclera. Long, sloping shoulders and well-defined withers; body deep and well sprung; mane and tail often sparse. Feet and limbs sound with good bone, hooves often striped. Action smooth and easy.

Haflinger Society of Great Britain

Address 13 Parkfield, Pucklechurch, Bristol BS17 3NR.

Breed Description Height (at three years old), 13.1 to 14.2hh (mares), and 13.3 to 14.2½hh (colts). Colour: chestnut – light, middle, liver or red; mane and tail flaxen. Head short with slight dish, large, dark and lively eyes, fine nostrils and small, pliable ears. Neck, strong and well positioned, not too short. Broad, deep chest, broad loins with good joints, a muscular croup that is not too short and a well-carried tail. Deep girth, measuring 65–75in. Clean limbs with hard, healthy hooves; strong forearms and a good second thigh; short cannons.

British Palomino Society

Address Penrhiwllan, Llandysul, Dyfed SA44 5NZ.

Breed Description The Palomino is not an individual breed, but a particular colour of horse represented by individuals from many different breeds. Colour: body colour is gold, ideally like a newly minted gold coin, but three shades lighter or darker are acceptable. White markings are permissible only on the legs and face. Mane and tail white, with not more than 15 per cent dark or chestnut hairs. Eyes, dark brown, hazel or black iris. Skin dark, except under white facial markings.

Appendix 2

Useful Addresses

GREAT BRITAIN

American Saddlebred Association of
Great Britain
Uplands
Alfriston
E. Sussex

British Andalusian Society
Church Farm
Church Street
Semington
Trowbridge
Wilts

British Bloodstock Agency Ltd
Thormanby House
Falmouth Avenue
Newmarket
Suffolk

British Friesian Horse Society
George and Dragon Hall
Mary Place
London

British Mule Society
Hope Mount Farm
Top of Hope
Alstonfield
Nr Ashbourne
Derbyshire

British Show Hack
Cob and Riding Horse Association
Rookwood
Packington Park
Meriden
Warwickshire

British Show Pony Society
124 Green End Road
Sawtry
Huntington
Cambs

British Warmblood Society
Moorlands Farm
New Yatt
Witney
Oxfordshire

Cleveland Bay Horse Society
York Livestock Centre
Murton
York

Equine Research Station
Balaton Lodge
Newmarket
Suffolk

Hunters Improvement and National
Light Horse Breeding Society
96 High Street
Edenbridge
Kent

National Foaling Bank
Meretown Stud
Newport
Shropshire

National Pony Society
Brook House
25 High Street, Alton
Hants

Ponies Association of UK
Chesham House
Green End Road
Sawtry
Huntingdon
Cambs

Shire Horse Society
East of England Showground
Peterborough

Thoroughbred Breeders Association
168 High Street
Newmarket
Suffolk

Weatherby and Son
42 Portman Square
London

USA AND CANADA

American Quarter Horse Association
27017 1-40
Box 200
Amarillo
TX 79168

American Saddlebred Horse
  Association
4093 Iron Works Pike
Lexington
KY 40511

American Shire Horse Association
Rt. 1
Box 10
Adel
IA 50003

American Standardbred Breeders
Box 11
Lexington
KY 40501

American Warmblood and Sports
  Horse Guild
Box 5202
Grants Pass
OR 97527

Appaloosa Horse Club
Box 840
Moscow
ID 83843

Arabian Horse Registry of America
12000 Zunie St.
Westminster
CO 80234

Clydesdale Breeders
17378 Kelley Road
Pecatonica
IL 61063

International Arabian Horse
  Association
(Half Arab and Anglo-Arab)
Box 33696
Denver
CO 80203

Jockey Club, The
380 Madison Avenue
New York
NY 10017

National Show Horse Registry, Inc.
10401 Linn Station Road, Suite 237
Louisville
KY 40223

Percheron Horse Association
Box 141
Fredericktown
OH 43019

Tennessee Walking Horse Breeders and
 Exhibitors
P.O. Box 286
Lewisburg
TN 37091

Canadian Arabian Registry
Box 101
Bnouden
Alta
TOM OKO

Canadian Equestrian Federation
333 River Road
Ottawa
Ont. KIL 8H9

Bibliography

Allen, W. Edward, *Fertility and Obstetrics in the Horse* (Blackwell Scientific Publications, 1988)

Houghton Brown, J. and Powell-Smith, V., *Horse and Stable Management* (Granada Publishing Ltd, 1984)

Loch, S., *The Royal Horse of Europe* (J. A. Allen)

Rees, L., *The Horse's Mind* (Stanley Paul, 1984)

Rossdale, P. D. and Ricketts, S. W., *Equine Stud Farm Medicine* (Bailliere Tindall, 1980)

Index